BASIC SKILLS PRACTICE MASTERS

Algebra 1

HOLT, RINEHART AND WINSTON

A Harcourt Classroom Education Company

Austin · New York · Orlando · Atlanta · San Francisco · Boston · Dallas · Toronto · London

To the Teacher

There are many basic mathematical skills that students must master in order to acquire ease and fluency in the handling of mathematical ideas. The *Basic Skills Practice* masters provide both instruction and practice in basic (pre-algebra) skills. There are three types of masters.

- Part 1 masters provide instruction, using an example-and-solution format, and practice exercises.

- Part 2 masters provide practice of the content taught in Part 1, using a standardized-test format.

- Cumulative Review masters provide test-format practice of all skills taught and practiced in preceding masters.

Photo Credit
Front Cover: (background), Index Stock Photography Inc./Ron Russell; (bottom), Jean Miele MCMXCII/The Stock Market.

Printed in the United States of America

ISBN 0-03-054277-4

4   5   6   7   066   02

Table of Contents

Basic Skills Practice

Metric and Customary Units, Part 1

You can convert between units of measure in either the customary system or the metric system. To convert between units of measure, first write an equation with 1 unit. Then find the conversion factor.

Customary System

1 mile (mi) = 1760 yards (yd)
1 mile (mi) = 5280 feet (ft)
1 yard = 3 feet
1 foot = 12 inches (in.)
1 gallon (gal) = 4 quarts (qt)
1 gallon = 128 ounces (oz)
1 quart = 2 pints (pt)
1 pint = 2 cups (c)
1 cup = 8 ounces

Metric System

1 kilometer (km) = 1000 meters (m)
1 meter = 100 centimeters (cm)
1 centimeter = 10 millimeters (mm)
1 liter (L) = 1000 milliliters (mL)
1 kilogram (kg) = 1000 grams (g)
1 gram = 1000 milligrams (mg)

Example 1:

5 ft = ___ in.
1 ft = 12 in.

Multiply each side by 5.

1 ft $\times$ 5 = 12 in. $\times$ 5

5 ft = 60 in.

Example 2:

70 cm = ___ m
100 cm = 1 m

Multiply each side by $\frac{70}{100}$.

$100 \text{ cm} \times \frac{70}{100} = 1 \text{ m} \times \frac{70}{100}$

70 cm = $\frac{7}{10}$ m, or 0.7 m

Convert the following by using the customary system:

1. 2 yd – _____ ft

2. 30 ft = _____ in.

3. 2640 ft = _____ mi

4. 3520 yd = _____ mi

5. 7 mi = _____ yd

6. 15 pt = _____ qt

7. 1536 oz = _____ gal

8. 96 qt = _____ gal

9. 144 ft = _____ yd

Convert the following by using the metric system:

10. 4 km = _____ m

11. 56 g = _____ kg

12. 133 L = _____ mL

13. 0.008 m = _____ cm

14. 70 cm = _____ m

15. 3200 mg = _____ g

16. 2300 mL = _____ L

17. 2 cm = _____ mm

18. 300 mm = _____ cm

Convert the following by using the appropriate system:

19. 37 m = _____ km

20. 31 pt = _____ c

21. 15,840 yd = _____ mi

22. 85 gal = _____ oz

23. 4.7 m = _____ cm

24. 13,200 ft = _____ mi

25. 75 m = _____ km

26. 2640 ft = _____ yd

27. 366 oz = _____ c

Basic Skills Practice

Metric and Customary Units, Part 2

Read each question and circle the best answer.

1. The garden club needs 275 feet of ribbon to make bows. The club president bought 102 yards of ribbon. How many **feet** of ribbon did the president buy?

    A 173 ft
    B 275 ft
    C 306 ft
    D 377 ft

2. Juan's aquarium holds 12 gallons of water. He is using a 1-quart pitcher to fill the aquarium. How many **quarts** of water will it take to fill the aquarium to capacity?

    F 3 qt
    G 13 qt
    H 36 qt
    I 48 qt

3. Joe bought 20 yards of rope. How many **feet** did he buy?

    A 240 ft
    B 100 ft
    C 60 ft
    D 20 ft

4. A race of 10,000 meters is being held on Saturday to benefit a local charity. How many **kilometers** will the race be?

    F 1 km
    G 5 km
    H 10 km
    I 100 km

5. Jordan has a pet snake that crawls into a tube 5400 millimeters long. How many **centimeters** long is the tube?

    A 54 cm
    B 64 cm
    C 320 cm
    D 540 cm

6. Martin is 6 feet 2 inches tall. How many **inches** tall is Martin?

    F 74 in.
    G 72 in.
    H 68 in.
    I 62 in.

Solve each problem.

7. Benjamin borrowed a device from his sister that measures how many yards he rides his bicycle. How many yards will the device show after he rides a total of 34 miles? _____

8. Wynona needs 42 pieces of ribbon, each 8 inches long. When she brings the spool of ribbon to the counter to be measured, the salesperson asks, "How many feet would you like?" What is Wynona's answer? _____

9. Calvin is using a measuring cup to fill his hummingbird feeder. He needs 2 pints of water. How many times does he fill his 1-cup measure? _____

Basic Skills Practice

Basic Skills Practice

Measurement and Precision, Part 1

Units of Measure		
	Customary	**Metric**
Length	miles (mi)	kilometers (km)
	yards (yd)	meters (m)
	feet (ft)	centimeters (cm)
	inches (in.)	millimeters (mm)
Volume	gallon (gal)	liter (L)
	quarts (qt)	milliliters or cubic centimeters (mL or cc)
	cups (c)	
Weight	pounds (lb)	kilograms (kg)
	ounces (oz)	grams (g)
		milligrams (mg)

The precision of a measuring instrument is the smallest division on the instrument.

The greatest precision for this ruler is $\frac{1}{4}$ in.

   yardstick

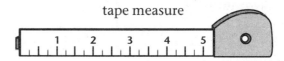   tape measure

1. What is the greatest precision for the yardstick? _____

2. What is the greatest precision for the tape measure? _____

Circle the most reasonable measure of length for each of the following:

3. the length of a teacher's desk   **a.** 90 cm   **b.** 90 km   **c.** 90 ft

4. the distance from Austin to San Antonio   **a.** 1000 km   **b.** 80 mi   **c.** 2000 ft

Circle the most reasonable measure of volume for each of the following:

5. the volume of a coffee pot   **a.** 1 kL   **b.** 50 c   **c.** 1.5 L

6. the volume of a jug of milk   **a.** 5 gal   **b.** 2 L   **c.** 300 mL

Circle the most reasonable measure of weight for each of the following:

7. the weight of a person   **a.** 185 lb   **b.** 400 kg   **c.** 200 oz

8. the weight of a feather   **a.** 1 lb   **b.** 80 oz   **c.** 10 mg

Basic Skills Practice

Measurement and Precision, Part 2

Read each question and circle the best answer.

1. Vanessa and her father are making a tree house. They need to cut a 4-yd board into 3 equal pieces. Which customary unit of measure will give the most precise measurement?

   A meters

   B centimeters

   C yards

   D inches

2.

   What is the greatest precision that can be reached with this ruler?

   F Nearest $\frac{1}{2}$ inch

   G Nearest $\frac{1}{4}$ inch

   H Nearest $\frac{1}{8}$ inch

   I Nearest $\frac{1}{16}$ inch

3. Which metric unit of measure would be the best choice for weighing very light objects, such as flower petals, a leaf, or a paper clip?

   A mm

   B mg

   C g

   D kg

4. Simone wants to measure the width of her book. Which is the best tool to use for measuring the width of her book?

   F yardstick

   G plastic ruler

   H 6-ft tape measure

   I 4-qt measuring cup

5. Juanita has a piece of string 2 meters long. Stephen has a piece of string 5 meters long. Gabriella has a piece of string that is the length of Juanita's string and Stephen's string put together. What is the length of Gabriella's string?

   A 7 liters

   B 5 yards

   C 7 ounces

   D 7 meters

6. Which metric unit of measure is more precise than centimeters for measuring the thickness of a textbook?

   F kg

   G kL

   H mm

   I km

Solve each problem.

7. A car holds 1.25 gallons of oil. If the car contains 3 quarts, how many *quarts* does the car need?

   _____

8. In the long drive golf contest, Mike's drive was 150 inches more than Abel's drive. How many more *feet* is this?

   _____

Basic Skills Practice
Estimation, Part 1

Use estimation to find *about how much*.

Estimating an answer before you work a problem helps you to spot errors.
You can estimate the size of your answer by rounding each number to the nearest easy number.
Then you can use the estimate to check calculations.

Example 1: 117.768 + 43.07 + 897.5
  100 + 50 + 900   ←Add the rounded numbers.
  1050   ←Approximate answer
  Your exact answer should be roughly the same as 1050. (The exact answer is
  1058.338.)

Example 2: 45.7 × 99.0357
  46 × 100   ←Round, and then multiply.
  4600   ←Approximate answer
  Your exact answer should be roughly the same as 4600. (The exact answer is
  4525.93149.)

Estimation is useful when an approximate answer is all that is needed.

Example 3: The restaurant bill for a meal came to a total of $23.75. You would like to leave a
  tip of about 15%. To the nearest dollar, how much should you leave?
  $23.75   ←Round to the nearest dollar.
  $24.00 × 0.15   ←Multiply rounded dollar amount by 15%.
  $3.60   ←Estimated amount of tip

Estimate approximate answers for each problem.

1. 3999 + 2107 + 33,022 _____

2. 8.72 − 3.34 _____

3. 73.94 + 16.3 _____

4. 59.33 − 23.31 _____

5. 45.7 × 99.0357 _____

6. 169.52 ÷ 46.8 _____

7. 1.23 × 9786 _____

8. (45.6 × 3.08) + 789.034 _____

Circle the most reasonable estimate for each.

9. the mass of a large pickle **a.** 70 kg **b.** 70 g **c.** 79 mg

10. the length of a dining room table **a.** 100 cm **b.** 100 km **c.** 100 m

11. the width of a room **a.** 4 m **b.** 4 mm **c.** 4 cm

12. the volume of a small fishbowl **a.** 1 kL **b.** 1 L **c.** 5 mL

Estimate an answer and then solve.

13. A sack of potting soil weighs almost 35 pounds. To the nearest
    thousand pounds, about how much will 785 sacks weigh? _____

Basic Skills Practice

Estimation, Part 2

Read each question and circle the best answer.

1. Sandra has $7.00. Each pack of gum costs $0.95 (tax included). How many packs of gum can Sandra purchase?

   A 6
   B 7
   C 8
   D 9

2. Bethany bought 3 pair of shoes priced at $19.95, $24.95, and $32.95. Which of the following would be a reasonable total for the 3 pairs, before tax is added?

   F less than $50
   G between $50 and $75
   H between $75 and $100
   I more than $100

3. Delmar is buying a beef roast that weighs 3.89 pounds. Estimate how much the roast will cost if it is selling for $2.09 per pound, not including tax?

   A $5
   B $6
   C $7
   D $8

4. If Sandra travels at an average speed of 52 miles per hour, approximately how many hours will it take her to drive 500 miles?

   F 6 hours
   G 8 hours
   H 10 hours
   I more than 10 hours

5. Sanborn took his younger brother and 3 of his brother's friends out for ice-cream cones. A single-dip cone is $0.95 with tax. About how much can Sanborn expect to spend on the 4 children and himself?

   A less than $3
   B between $3 and $4
   C between $4 and $6
   D more than $6

6. The area of a rectangular lot is 7230 square feet. If the length of the lot is about 120.5 feet, which is the best estimate for the width of the field?

   F 100
   G 90
   H 70
   I 60

Solve each problem.

Nathan and Samantha are choosing from the gift list shown at right. Estimate the cost of the following purchases. Assume that tax is included in the prices. Round the cost **up** to the nearest dollar.

7. blender and silver platter _____

8. bowls and silver platter _____

9. pizza pan and blender _____

Gift List	
blender...$30.35
pizza pan with cutter $12.95
set of bowls $18.25
silver platter...............................$15.65

Basic Skills Practice

Order of Operations, Part 1

When you have more than one operation in one problem, you need to know the rule that tells which operations are done first so that you get consistent answers.

Step 1: First, do all operations in parentheses.
$18 - 9 \div 3 + 2^2 \times (7 - 2)$
$18 - 9 \div 3 + 2^2 \times 5$

Step 2: Simplify exponents.
$18 - 9 \div 3 + 2^2 \times 5$
$18 - 9 \div 3 + (2 \times 2) \times 5$
$18 - 9 \div 3 + 4 \times 5$

Step 3: Do all multiplications and divisions in order from left to right.
$18 - 9 \div 3 + 4 \times 5$
$18 - 3 + 20$

Step 4: Do all additions and subtractions in order from left to right.
$18 - 3 + 20$
$15 + 20$
35

The sentence "*Please excuse my Dear Aunt Sally*" might help you remember the correct order of operations. Notice that the first letter of each word also is the first letter of an operation (*p* for *parentheses, e* for *exponents, m* for *multiplication, d* for *division, a* for *addition,* and *s* for *subtraction*). This is called a mnemonic device.

Simplify each expression. Show each step.

1. $2 + 6 \times 8 \div 4$ _____

2. $10 \div 5 \times 2 + 6$ _____

3. $(4 + 20) \times 3$ _____

4. $4 + (20 \times 3)$ _____

5. $30 - 2^2 + (5 + 5)$ _____

6. $(8 - 5)^2 \times 2 + 5$ _____

7. $21 + 16 + 7 - 6 \div 2$ _____

8. $3 + (2 \times 7) \times 10$ _____

9. $3 + (6 \times 7) \div 2$ _____

10. $1 + 30 - (4 \times 5)$ _____

11. $6(2 + 5) - 15$ _____

12. $-2(5 - 8) + 18 \div 3$ _____

13. $5 \times 3 - 4 \div 2 + 9$ _____

14. $4^2 \times 2 \div 4 - 5$ _____

15. $(4 + 8 - 2) \times 15$ _____

16. $22 + 5 \times 2 - 5$ _____

17. $(7 + 5) \times (8 - 4)$ _____

18. $5 + 8 \div 8 \times 2$ _____

19. $5^2 - 64 \div 4$ _____

20. $10^2 \div 10 + 8 \times 4$ _____

Basic Skills Practice

Order of Operations, Part 2

Read each question and circle the best answer.

1. Ahmed bought a shirt that was half off the regular price of $24.98. When he got to the counter, the salesperson took another $5.00 off. Use the correct order of operations to determine the final price of the shirt.
$$\$24.98 - 0.5 \times \$24.98 - \$5.00$$

   A $7.49
   B $9.99
   C $12.49
   D $14.99

2. Which one of the following is true?

   F $12 \div (3 + 1) = 5$
   G $33 - 54 \div 6 = 24$
   H $47 - 5 \times 2 = 84$
   I $7 \times (3 + 15) = 36$

3. Simplify the following expression:
$$30 \div 5 \times 2 + 5 - 4^2$$

   A 0
   B 1
   C 8
   D 24

4. The expression $4(3x + y)$ is equivalent to—

   F $3(4x + y)$
   G $4 \cdot 3x + y$
   H $4 \cdot 3x + 4 \cdot 3y$
   I $4 \cdot 3x + 4y$

5. Choose the expression that shows the correct placement of the parentheses to make the following equation true:
$$41 - 45 \div 9 \times 2 = 31$$

   A $41 - [(45 \div 9) \times 2]$
   B $41 - (45 \div (9 \times 2))$
   C $[41 - (45 \div 9)] \times 2$
   D $(41 - 45) \div (9 \times 2)$

6. Simplify the following expression:
$$(90 \div 3^2) - 9 + 5^2$$

   F 24
   G 26
   H 98
   I 126

Solve each problem.

7. Calculate: $(4 - 2)^2 \times 7$. _____

8. Calculate: $7(1 + 8) \div 9$. _____

9. Calculate: $66 - 21 \times 4 + 87$. _____

10. Calculate: $3^2 - 11 + 2(6 \times 2)$. _____

11. Calculate: $28 + (0 + 4) - (10 \div 2)$. _____

12. Calculate: $33 \div (3 \times 11) + 72$. _____

13. Calculate: $(12 - 4 \div 2) \times 5$. _____

Basic Skills Practice

Basic Skills Practice

Cumulative Review

Read each question and circle the best answer.

1. Robert and his father are buying some items for a camping trip. These items are listed below. Estimate the amount of money they spent (not including tax).

    2 sleeping bags ....$25.95 each
    1 gas grill $42.99
    bug spray $2.27
    1 pair of sunglasses...............$12.99

    A $81
    B $84
    C $107
    D $110

2. Calculate the following expression:
    $8 + 2^2 - (2 \times 4 - 8)$

    F 0
    G 12
    H 18
    I 24

3. Gillian is measuring a large sheet of paper for a mural. She wants the paper to be $5\frac{1}{2}$ ft by 8 ft. When she locates her yardstick, she finds that her brother has broken it. She must measure the paper in inches. What size should it be in inches?

    A 60 in. by 96 in.
    B $16\frac{1}{2}$ in. by 24 in.
    C 55 in. by 80 in.
    D 66 in. by 96 in.

4. Mateo and Cecil want to determine who can shotput the farthest. Which of the following is the best unit of measure to use?

    F km
    G m
    H cm
    I in.

Solve each problem.

5. Calculate: $(15 - 10)^2 + 15 - 5^2$. _____

6. How many milliliters are in 5.256 L? _____

7. A bookshelf is 3 feet wide. How many inches wide is it? _____

8. About how many bows could you make out of 275 ft of ribbon if each bow takes 4.75 ft? _____

9. Guidardo ran a 10-kilometer marathon. How many meters did Guidardo run? _____

10. What is the precision of the ruler shown? _____

Basic Skills Practice

Evaluating Expressions and Formulas, Part 1

To evaluate an expression or a formula, rewrite it, and replace the variables with the given values. Then simplify by using the order of operations.

Example 1: Evaluate $\dfrac{3x + z^2}{y + 1}$ when $x = 4$, $y = 23$, and $z = 6$.

First rewrite with the given values. $\quad \dfrac{3(4) + (6)^2}{23 + 1}$

Then simplify. $\quad \dfrac{12 + 36}{24} = \dfrac{48}{24} = 2$

Example 2: The formula for changing degrees Celsius to degrees Fahrenheit is $F = \dfrac{9}{5}C + 32$, where C is the temperature in degrees Celsius and F is the temperature in degrees Fahrenheit.

Find the Fahrenheit temperature if the Celsius temperature is 35°C.

First replace C with the given value. $\quad F = \dfrac{9}{5}(35) + 32$

Then simplify. $\quad F = 63 + 32$
$\quad F = 95°\text{F}$

Evaluate when $x = 3$.

1. $x + 3^2$ _____

2. $(2 + x)^2$ _____

3. $(x + 7) \times 3$ _____

4. $420 \div (60 \div x)$ _____

5. $50 - x^3 \times 5$ _____

6. $(13 - x)^2 + 5$ _____

7. $(9 \div x)^2 - 9$ _____

8. $3 + [(13 - x) \times 21]$ _____

Evaluate when $y = 1$ and $z = 2$.

9. $(3 + y) \div 2 - z$ _____

10. $12 - (z - y)^2$ _____

11. $\dfrac{z + y^3}{3}$ _____

12. $\dfrac{y}{z + 3y}$ _____

13. $\left(\dfrac{z + 2}{y + 1}\right)^2 \div 4$ _____

14. $(z \times y) - (4 + y)$ _____

15. $4 \times 3 - 5 + (2 \times z)^2$ _____

16. $\dfrac{(4 + y)^2}{5} + z$ _____

Use the Celsius temperature given to find the Fahrenheit temperature.

17. Celsius = 10°C Fahrenheit = _____

18. Celsius = 25°C Fahrenheit = _____

19. Celsius = 100°C Fahrenheit = _____

20. Celsius = 0°C Fahrenheit = _____

21. Celsius = 37°C Fahrenheit = _____

22. Celsius = 50°C Fahrenheit = _____

Basic Skills Practice

Evaluating Expressions and Formulas, Part 2

Read each question and circle the best answer.

1. The total area of 6 identical square pieces of cloth is $A = 6l^2$, where l is the length of the side of the square. When l is 2 cm, what is the total area for 6 of the squares, in square centimeters?

   A 0

   B 12

   C 18

   D 24

2. Ashley had $127.34 in her savings account. After she withdrew $48.65, how much remained?

   F $78.69

   G $78.71

   H $81.31

   I $175.99

3. An aquarium is 2 feet wide, 4 feet long, and 2 feet high. What is the volume of the aquarium?

   A 7 ft^3

   B 8 ft^3

   C 12 ft^3

   D 16 ft^3

4. Susan bought tickets to a concert for herself and 9 friends. A ticket costs $14.75, but a discount of $2.00 per ticket is given when 10 or more tickets are purchased. Susan used this formula to determine the total cost of tickets: $T = \$14.75(x) - \$2.00(x)$, where x is the number of tickets purchased. Find the total cost of the tickets if $x = 10$.

   F $127.50

   G $204

   H $220

   I $234

5. What is the area of a rectangle if $l = 8$ and $w = 3$?

   A 11

   B 17

   C 22

   D 24

6. The cost of taking a taxi is given by the formula, $C = \$2.00 + 0.50m$, where m is the number of miles. What is the total cost if the trip is 15 miles long?

   F $8.00

   G $9.50

   H $10.50

   I $11.50

Solve each problem.

7. Evaluate $(2x - 15) \times 3$ when $x = 3$. _____

8. Evaluate $x^2 + 2(x - 9)$ when $x = 5$. _____

9. Evaluate $3(x \div 7) + 51$ when $x = 49$. _____

10. Evaluate $(x + 3) \div 5 + 33$ when $x = 17$. _____

11. Evaluate $[(42 - x) \div 10] \times 15$ when $x = 12$. _____

Basic Skills Practice
Finding Patterns, Part 1

One way you can find a pattern in a sequence is by using addition.

Example 1: Find the next two numbers in the sequence 6, 13, 20, 27, 34, _____, _____.

6　＋7　13　＋7　20　＋7　27　＋7　34　＋7　?　＋7　?

By using addition, you find that adding 7 to the previous number gives the next number in the sequence.

Thus, 34 + 7 = 41 and 41 + 7 = 48. The next two numbers in the sequence are 41 and 48.

Another way to find a pattern is to look for what stays the same. Then find what changes.

Example 2: What is the next expression in the pattern *aBc, dEf, gHi,* _____?

Looking at each of the expressions you see that the middle letter is always a capital, the others are not. If you then read the pattern aloud, "abcdefghi," you hear the alphabetical pattern that allows you to find what changes.

Find the next number of each sequence.

1. 2, 4, 6, 8, _____
2. 1, 4, 7, 10, _____
3. 9, 18, 27, 36, _____

4. 1, 6, 11, 16, _____
5. 10, 20, 30, _____
6. 3, 7, 11, 15, _____

7. 7, 14, 21, 28, _____
8. 6, 15, 24, 33, _____
9. 1, 2, 3, 4, 5, _____

Fill in the missing element for each sequence.

10. 212, 424, _____, 848, 1060
11. 11, 22, 33, _____, 55

12. 121, 232, _____, 454, 565
13. AaA, bBb, CcC, _____, EeE, fFf

14. AcE, FhJ, KmO, _____
15. A1a, B2b, C3c, _____, E5e, F6f

16. 246, 369, 4812, 51,015, _____
17. 300, 600, 400, 700, _____, 800

18. 555, 444, 333, _____
19. 145, 256, 367, _____

Find the missing element(s) in each sequence.

20. 13, 26, 39, _____, 65
21. 26, 35, _____, 53, 62

22. Abd, Efh, Ijl, _____, _____
23. 123, 456, 789, 012, 345, _____

24. 543, 654, _____, 876, 987
25. 12D, 23E, 34F, 45G, _____

26. AZ, BY, CX, _____, _____
27. 3, 9, 27, _____, 243, _____

28. Z0A, Y1B, X2C, _____
29. 15, 30, 60, 120, _____

Basic Skills Practice

Finding Patterns, Part 2

Read each question and circle the best answer.

1. What number is next in this pattern?
   15, 21, 27, 33, 39, . . .

   A 44
   B 45
   C 40
   D 39

2. Phillipe is paid an allowance each month based on the number of chores he performs. Use the table to find how much allowance he would get paid if he did 11 chores this month.

# Chores	Allowance
5	$3.00
7	$6.00
9	$9.00

   F $9.00
   G $12.00
   H $18.00
   I $10.00

3. Find the missing number in this pattern.
   5, 10, _____, 20, 25

   A 11
   B 15
   C 21
   D 30

4. Find the missing number in this pattern.
   111, 333, _____, 777

   F 555
   G 444
   H 666
   I 222

5. Vanessa is designing a bracelet. Cecily is trying to learn the pattern. Which type of bead should Cecily add next?

   A ⬭
   B ●
   C ○
   D ●●

6. A train leaves the Marshall Station for Chicago at regular intervals. It left for Chicago at 11:10 A.M, 11:45 A.M, and 12:20 P.M. When is the next departure leaving for Chicago?

   F 12:25 P.M.
   G 12:45 P.M.
   H 12:55 P.M.
   I 1:05 P.M.

7. Find the missing number in this pattern.
   232, 343, _____, 565, 676, 787

   A 424
   B 403
   C 454
   D 554

8. Find the missing number in this pattern.
   9, _____, 49, 69, 89

   F 29
   G 19
   H 39
   I 27

Basic Skills Practice

Using Differences to Find Patterns, Part 1

Some patterns can be understood by using differences.

Example 1: Find the next number in the sequence 5, 25, 43, 59, 73, …

First: Find the differences between each of the pairs in the sequence.

5 — 25 — 43 — 59 — 73
  20   18   16   14

These differences are not the same.

Next: Since the first differences are not constant, we look at the second differences.

20 — 18 — 16 — 14
  −2   −2   −2

These differences are constant.

Finally: Draw the pattern in this way:

5 — 25 — 43 — 59 — 73 — 85
 +20 20−2 20−2−2 20−2−2−2 20−2−2−2−2
    +18   +16   +14   +12

The number after 73 in the sequence is 85.

Example 2: You may also find patterns by using multiplication.
Find the next number in the sequence 7, 21, 63, 189, …

Think: 7×3 is 21, 21×3 is 63, and 63×3 is 189. The pattern is 3 multiplied by the previous number. The next number must be 189×3, or 567.

Find the next number of each sequence.

1. 0, 8, 19, 33, 50, 70, _____

2. 0, 71, 133, 186, _____

3. 5, 55, 100, 140, 175, 205, _____

4. 223, 220, 210, 193, 169, _____

5. 12, 18, 24, 30, 36, _____

6. 3, 1, −1, −3, −5, _____

Fill in the missing number for each sequence.

7. 3, 6, 12, 24, 48, _____

8. 448, 224, 112, _____, 28, 14, 7

9. 1250, 250, 50, _____, 2

10. 11, 44, _____, 704

11. 6, 10, 16, 24, 34, _____

12. 9, 12, 15, _____, 21

Find the missing number.

13. 1, 7, 49, _____, 2401

14. 111, 166, _____, 243, 265, 276

15. 1203, 1198, 1191, _____, 1171

16. 256, 192, 144, _____, 96

17. 11, 6, 1, −4, _____

18. 21, 42, 84, _____, 336

Basic Skills Practice

Basic Skills Practice

Using Differences to Find Patterns, Part 2

Read each question and circle the best answer.

1. Suppose you begin the school year with 1000 sheets of notebook paper. If you are using 6 sheets a day, after how many days will you have only 100 sheets left?

   A 175
   B 170
   C 150
   D 140

2. Estelle is cutting fabric pieces for a quilt. She cuts 8 blue and 24 navy hexagons. Then she cuts a number of light green hexagons and 96 dark green hexagons. Assuming that the ratio of light green to dark green is the same as that of blue to navy, how many squares of light green did she cut?

   F 4
   G 16
   H 48
   I 32

3. Find the missing number in the pattern.
   5724, 1908, _____, 212

   A 418
   B 542
   C 1048
   D 636

4. Find the missing number in the pattern.
   7, 30, 57, 88, _____, 162, 205

   F 112
   G 100
   H 123
   I 99

5. The number of court cases in a city was 733 in 1990 and has been increasing by about 15 cases per year since then. If this pattern continues, in what year will the number of cases exceed 1000?

   A 1998
   B 2002
   C 2006
   D 2008

6. Bruce is taking care of his friend's plants for the next 4 weeks. After the first week, 81 of the original 87 plants are still alive. After the second week, 76 are alive. After the third week, 72 are alive. How many plants will still be alive when the friend returns home?

   F 64
   G 66
   H 69
   I 70

7. Find the missing number in the pattern.
   13, 39, _____, 351

   A 62
   B 117
   C 91
   D 252

8. Find the missing number in the pattern.
   _____, 7, 17, 36, 64

   F −2
   G 1
   H 0
   I 6

Basic Skills Practice

Cumulative Review

Read each question and circle the best answer.

1. A kitchen counter is 36 inches above the floor. How many **feet** high must the stove be to match the height of the counter?

  A 1 ft

  B 3 ft

  C 3 ft 6 in.

  D 3.6 ft

2. $4(3 + 2) - \dfrac{10}{2} =$

  F 16

  G 15

  H 7

  I 4

3. Find the value of the following expression when $x = 2$ and $y = 0$:
$$5 + x^2 y$$

  A 0

  B 2

  C 5

  D 9

4. Find the missing number in the following sequence:
$$121, 232, 343, 454, \_\_\_\_, 676, 787$$

  F 554

  G 555

  H 565

  I 656

5. In a science-fiction movie, space aliens are taking humans from Napersville on a daily basis. The town initially had 3328 people. At the end of the first day, 1664 people are left. After two days 832 people are left. How many people will be left in Napersville at the end of the third day?

  A 0

  B 128

  C 312

  D 416

6. Kaitlin is learning to read. She can read 15 words at the end of a week. By the end of the second week, she can read 30 words. After three weeks she can read 45 words. If this pattern continues, how many words will she be able to read at the end of the fourth week?

  F 45 words

  G 55 words

  H 60 words

  I 75 words

7. Find the missing number in the following sequence:
$$1, 5, 10, 16, \_\_\_\_, 31$$

  A 23

  B 22

  C 20

  D 25

Solve each problem.

8. Evaluate $z^2 w - 5$ when $w = 3$ and $z = 2$.   _____

9. Evaluate $z^2 + wz$ when $w = 3$ and $z = 2$.   _____

10. What is the missing number in the sequence $9, 3, 1, \dfrac{1}{3}, \_\_\_\_, \dfrac{1}{27}$?   _____

Basic Skills Practice

Fractions and Equivalent Fractions, Part 1

Fractions, as well as percentages and decimals, indicate a part of a whole.

Example: The picture represents a granola bar that
Jennifer bought. She gives half to Michael.
Michael gives half of his share to Raoul.
How much of the granola bar does Raoul have?
Divide the granola bar to represent each person's portion.

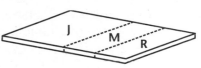

Because the portions of the granola bar are **not equal,** divide Jennifer's portion into
2 equal parts.
The parts are now equal, so you can find Raoul's
portion.

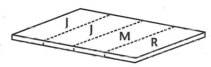

$$\text{Raoul's portion} = \frac{\text{Raoul's part}}{\text{Number of equal parts in the whole}}$$

Raoul has $\frac{1}{4}$ of the granola bar.

Write a fraction to show the amount of each shaded region.

1.    _____

2. 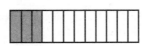   _____

3.   _____

4.   _____

5. 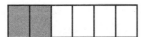   _____

6.    _____

**Write = for the fractions that are equivalent and ≠ for fractions
that are not equivalent.**

7. $\frac{2}{3}$ _____ $\frac{14}{21}$

8. $\frac{1}{9}$ _____ $\frac{2}{16}$

9. $\frac{5}{8}$ _____ $\frac{25}{80}$

10. $\frac{3}{11}$ _____ $\frac{12}{44}$

11. $\frac{12}{18}$ _____ $\frac{4}{6}$

12. $\frac{7}{9}$ _____ $\frac{63}{81}$

Complete each equivalent fraction.

13. $\frac{5}{25} = \frac{?}{5}$ _____

14. $\frac{2}{7} = \frac{?}{49}$ _____

15. $\frac{18}{66} = \frac{?}{11}$ _____

16. $\frac{3}{17} = \frac{?}{68}$ _____

17. $\frac{7}{56} = \frac{?}{8}$ _____

18. $\frac{4}{19} = \frac{?}{57}$ _____

19. $\frac{22}{48} = \frac{33}{?}$ _____

20. $\frac{9}{51} = \frac{15}{?}$ _____

21. $\frac{14}{4} = \frac{21}{?}$ _____

Basic Skills Practice

Fractions and Equivalent Fractions, Part 2

Read each question and circle the best answer.

1. In the following list, which fraction is *not* equivalent to the other fractions?

$$\frac{3}{7}, \frac{12}{28}, \frac{3}{14}, \frac{18}{42}, \frac{9}{21}$$

    A $\frac{3}{14}$

    B $\frac{9}{21}$

    C $\frac{12}{28}$

    D $\frac{18}{42}$

2. Yasmine needs $\frac{3}{4}$ of a cup of sugar for a recipe that she is making. She wants to use her $\frac{1}{8}$-cup measuring container. How many times does she need to fill this container to get the correct amount of sugar?

    F 7

    G 6

    H 5

    I 4

3. In which of the following diagrams does the shaded region represent $\frac{5}{9}$?

    A

    B

    C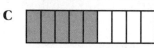

    D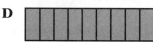

4. Josh is measuring pieces of wood for a CD holder that he is building. One of the pieces needs to be $\frac{14}{16}$ in. The smallest unit on Josh's ruler is $\frac{1}{8}$ in. What should the length of the wood be using this unit of measure?

    F $\frac{1}{8}$ in.

    G $\frac{2}{8}$ in.

    H $\frac{7}{8}$ in.

    I $\frac{13}{8}$ in.

Solve each problem.

5. The ratio of girls to boys is 7 to 9. If there are 35 girls, how many boys are there? _____

6. Jill drives 90 miles in 2 hours. How many miles can she travel in 3 hours? _____

7. Thirteen out of 15 students completed the test before the end of the class period. If there are 45 students in the class, how many finished the test? _____

Basic Skills Practice

Comparing Fractions, Part 1

You can compare two fractions that have the *same* denominator by comparing their numerators.

Example: Compare $\frac{3}{8}$ and $\frac{5}{8}$. Since $5 > 3$, $\frac{5}{8} > \frac{3}{8}$. To see this, use a picture of each fraction.

$\frac{5}{8}$ ▢▢▢▢▢▢▢▢ $\frac{3}{8}$ ▢▢▢▢▢▢▢▢ $\frac{5}{8} > \frac{3}{8}$

You can compare two fractions that have *different* denominators by using several methods.

Method 1: Shade drawings and compare visually. Compare $\frac{1}{2}$ and $\frac{2}{5}$.

$\frac{1}{2}$ $\frac{2}{5}$ $\frac{1}{2} > \frac{2}{5}$

Method 2: To compare two fractions with the same numerator, compare denominators. Compare $\frac{3}{7}$ and $\frac{3}{8}$. Remember that if there are *fewer* divisions, each division is larger. So $\frac{3}{7} > \frac{3}{8}$.

Method 3: Rewrite the fractions with a common denominator. Compare $\frac{2}{3}$ and $\frac{3}{5}$.

Multiply the denominators to find a common denominator: $3 \times 5 = 15$.

$\frac{2}{3} \times \frac{5}{5} = \frac{10}{15}$ $\frac{3}{5} \times \frac{3}{3} = \frac{9}{15}$ $\frac{10}{15} > \frac{9}{15}$ So $\frac{2}{3} > \frac{3}{5}$.

Shade the portion that each fraction represents. Then compare the fractions.

1. $\frac{1}{5}$  $\frac{2}{5}$ $\frac{1}{5}$ _____ $\frac{2}{5}$

2. $\frac{7}{10}$  $\frac{5}{12}$ $\frac{7}{10}$ _____ $\frac{5}{12}$

3. $\frac{5}{14}$ $\frac{3}{5}$ $\frac{5}{14}$ _____ $\frac{3}{5}$

Write =, < or > to compare each pair of fractions.

4. $\frac{2}{20}$ _____ $\frac{1}{5}$ 5. $\frac{6}{15}$ _____ $\frac{2}{5}$ 6. $\frac{18}{24}$ _____ $\frac{7}{8}$ 7. $\frac{3}{5}$ _____ $\frac{7}{10}$

8. $\frac{2}{3}$ _____ $\frac{5}{6}$ 9. $\frac{4}{17}$ _____ $\frac{3}{17}$ 10. $\frac{3}{4}$ _____ $\frac{13}{16}$ 11. $\frac{16}{19}$ _____ $\frac{80}{95}$

Basic Skills Practice

Comparing Fractions, Part 2

Read each question and circle the best answer.

1. On a fishing trip, Marcello caught 4 catfish. The fish weighed $3\frac{1}{3}$ lb, $3\frac{1}{4}$ lb, $3\frac{2}{3}$ lb, and $3\frac{5}{6}$ lb. Which group of numbers shows the sizes of the fish in the correct order from least to greatest?

   A $3\frac{1}{4}, 3\frac{1}{3}, 3\frac{2}{3}, 3\frac{5}{6}$

   B $3\frac{1}{3}, 3\frac{1}{4}, 3\frac{2}{3}, 3\frac{5}{6}$

   C $3\frac{1}{3}, 3\frac{2}{3}, 3\frac{1}{4}, 3\frac{5}{6}$

   D $3\frac{1}{3}, 3\frac{2}{3}, 3\frac{5}{6}, 3\frac{1}{4}$

2. Donny, Sera, and Casey each have a candy bar that is the same size. Donny eats $\frac{1}{2}$ of his candy bar, Sera eats $\frac{2}{3}$ of her candy bar, and Casey eats $\frac{3}{4}$ of his candy bar. Who ate the most?

   F Donny
   G Sera
   H Casey
   I They all ate the same amount.

3. Beverly is sorting bolts. She wants to start with the smallest and end with the largest. Which shows the correct order?

   A $\frac{3}{16}, \frac{1}{4}, \frac{5}{16}, \frac{1}{2}$

   B $\frac{1}{4}, \frac{3}{16}, \frac{5}{16}, \frac{1}{2}$

   C $\frac{1}{4}, \frac{1}{2}, \frac{3}{16}, \frac{5}{16}$

   D $\frac{1}{2}, \frac{5}{16}, \frac{1}{4}, \frac{3}{16}$

4. Richard is buying cord for a project. The cord is available in diameters of $\frac{5}{16}, \frac{1}{2}, \frac{3}{4}$, and $\frac{5}{8}$ in. If he wants the **_thickest_** cord, which diameter should he choose?

   F $\frac{1}{2}$ in.

   G $\frac{3}{4}$ in.

   H $\frac{5}{8}$ in.

   I $\frac{5}{16}$ in.

Solve each problem.

5. Janet wrote $\frac{3}{4}$ of her term paper in 6 days. Rebecca wrote $\frac{3}{8}$ of her paper in the same amount of time. Write the fraction for the student who is almost finished writing her paper. _____

6. Pam bought $\frac{9}{16}$ of a yard of fabric. She needs $\frac{5}{8}$ of a yard. How much more fabric does she need? _____

Basic Skills Practice

Mixed Numbers, Part 1

Use mixed numbers when you have one (or more) whole units along with part of another unit.

Example:

$$1 + 1 + \frac{1}{3} = 2\frac{1}{3}$$

Rewrite mixed numbers as improper fractions by dividing each of the whole units into parts to match the partial unit. Then add all of the parts together.

$$\frac{3}{3} + \frac{3}{3} + \frac{1}{3} = \frac{7}{3}$$

Here is a shortcut way to rewrite a mixed number as an improper fraction.

Step 1: Multiply the denominator by the whole number.

Step 2: Add the numerator to the product.

Step 3: Write the sum over the original denominator.

$$2\frac{1}{3} = \frac{(2 \times 3) + 1}{3} = \frac{7}{3}$$

Write each mixed number as an improper fraction.

1. $1\frac{2}{3}$ _____

2. $2\frac{1}{4}$ _____

3. $5\frac{1}{3}$ _____

4. $3\frac{1}{2}$ _____

5. $9\frac{3}{7}$ _____

6. $2\frac{5}{7}$ _____

7. $11\frac{5}{9}$ _____

8. $5\frac{2}{5}$ _____

9. $6\frac{1}{12}$ _____

10. $4\frac{3}{8}$ _____

11. $7\frac{5}{6}$ _____

12. $8\frac{2}{11}$ _____

Write each fraction as a whole number or mixed number in simplest form.

13. $\frac{22}{8}$ _____

14. $\frac{13}{6}$ _____

15. $\frac{43}{9}$ _____

16. $\frac{32}{7}$ _____

17. $\frac{24}{12}$ _____

18. $\frac{66}{8}$ _____

19. $\frac{14}{4}$ _____

20. $\frac{11}{2}$ _____

21. $\frac{27}{5}$ _____

22. $\frac{63}{9}$ _____

23. $\frac{56}{3}$ _____

24. $\frac{81}{4}$ _____

Basic Skills Practice

Mixed Numbers, Part 2

Read each question and circle the best answer.

1. Rachel is buying a new tennis racket. The grips have sizes of $4\frac{1}{2}$ in., $4\frac{3}{16}$ in., $4\frac{5}{8}$ in., and $4\frac{1}{4}$ in. Rachel wants the racket with the largest grip. Which does she buy?

   A $4\frac{3}{16}$ in. grip

   B $4\frac{1}{4}$ in. grip

   C $4\frac{1}{2}$ in. grip

   D $4\frac{5}{8}$ in. grip

2. Paul is making a cookie recipe that calls for $2\frac{2}{3}$ cups flour. He uses a $\frac{1}{3}$-cup measuring container. How many times must he fill the container to get the correct amount of flour?

   F 4
   G 6
   H 7
   I 8

3. Joan wants to place wrenches on a shelf in the order of their size. Which group of numbers shows the correct way to order the wrenches from largest to smallest?

   A $\frac{3}{8}$ in., $\frac{1}{2}$ in., $\frac{3}{4}$ in., $\frac{5}{8}$ in., $\frac{15}{16}$ in.

   B $\frac{5}{8}$ in., $\frac{3}{8}$ in., $\frac{15}{16}$ in., $\frac{3}{4}$ in., $\frac{1}{2}$ in.

   C $\frac{15}{16}$ in., $\frac{5}{8}$ in., $\frac{3}{4}$ in., $\frac{1}{2}$ in., $\frac{3}{8}$ in.

   D $\frac{15}{16}$ in., $\frac{3}{4}$ in., $\frac{5}{8}$ in., $\frac{1}{2}$ in., $\frac{3}{8}$ in.

4. The mixed numbers $3\frac{2}{3}$, $3\frac{7}{9}$, and $3\frac{5}{6}$ are in order from least to greatest. If the mixed number $3\frac{5}{9}$ is to fit into the order, where should it be placed?

   F Before $3\frac{2}{3}$

   G Between $3\frac{2}{3}$ and $3\frac{7}{9}$

   H Between $3\frac{7}{9}$ and $3\frac{5}{6}$

   I After $3\frac{5}{6}$

Solve each problem.

5. Mr. Daniels bought 75 shares of stock for $9\frac{3}{4}$ per share. After it increased $1\frac{3}{4}$ per share, he sold it. What was the price of the 75 shares of stock when he sold it? _____

6. Rick is $6\frac{1}{4}$ feet tall. If John is only $5\frac{7}{8}$ feet tall, how many feet taller is Rick? _____

7. Maxwell works $7\frac{1}{2}$ hours per day, 5 days per week, and earns $9.00 per hour. What is his weekly salary? _____

Basic Skills Practice

Cumulative Review

Read each question and circle the best answer.

1. Tiffany needs $2\frac{3}{4}$ gallons of paint for her room, Marcus needs $2\frac{1}{5}$ gallons, and Lucinda needs $2\frac{1}{3}$ gallons. Who needs the most paint?

    A Tiffany
    B Marcus
    C Lucinda
    D They all need the same amount.

2. Kathy's slice of pie was $\frac{1}{8}$ of the pie. Samuel's slice was $\frac{3}{16}$ of the pie, and Jackson's slice was $\frac{1}{4}$ of the pie. Who had the biggest slice?

    F Kathy
    G Samuel
    H Jackson
    I They each had the same amount.

3. Which of the following fractions is not equivalent to the others?
    $$\frac{9}{11}, \frac{81}{99}, \frac{27}{33}, \frac{37}{44}$$

    A $\frac{9}{11}$
    B $\frac{81}{99}$
    C $\frac{27}{33}$
    D $\frac{37}{44}$

4. Jorge is sorting pipe by diameter from smallest to largest. The various pipe diameters are $\frac{3}{8}$ in., $\frac{1}{2}$ in., $\frac{1}{4}$ in., and $\frac{3}{4}$ in. What is the correct order?

    F $\frac{1}{2}$ in., $\frac{1}{4}$ in., $\frac{3}{4}$ in., $\frac{3}{8}$ in.
    G $\frac{3}{4}$ in., $\frac{1}{4}$ in., $\frac{1}{2}$ in., $\frac{3}{8}$ in.
    H $\frac{1}{4}$ in., $\frac{3}{8}$ in., $\frac{1}{2}$ in., $\frac{3}{4}$ in.
    I $\frac{3}{8}$ in., $\frac{1}{4}$ in., $\frac{1}{2}$ in., $\frac{3}{4}$ in.

Solve each problem.

5. Ryan bought a shirt that was 30% off the regular price of $36.00. When he got to the counter, the salesperson took another $2.50 off. What is the final price that Ryan paid for the shirt? (Without tax) _____

6. The total area of 5 identical square pieces of cloth is $A = 5l^2$ where l is the length of the side of the square. When l is 4 cm, what is the total area for the 5 squares? _____

Basic Skills Practice

Adding and Subtracting Fractions, Part 1

To add or subtract fractions with a common denominator, add only the numerators.

Example 1: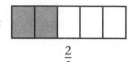

$$\frac{2}{5} \quad + \quad \frac{1}{5} \quad = \quad \frac{3}{5}$$

To add or subtract fractions with different denominators, write as equivalent fractions with common denominators. Then perform the appropriate operation.

Example 2: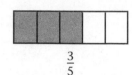

$$\frac{3}{4} \quad - \quad \frac{3}{8} \quad = \quad \frac{6}{8} \quad - \quad \frac{3}{8} \quad = \quad \frac{3}{8}$$

To add or subtract mixed numbers, write as improper fractions. Then perform the appropriate operation.

Add. Write the sum in simplest form.

1. $\frac{2}{9} + \frac{5}{9}$ _____

2. $1\frac{7}{10} + 4\frac{2}{5}$ _____

3. $1\frac{2}{3} + \frac{11}{15}$ _____

4. $\frac{5}{8} + \frac{3}{4}$ _____

5. $\frac{1}{7} + \frac{3}{5}$ _____

6. $1\frac{5}{16} + 2\frac{13}{16}$ _____

7. $5\frac{2}{3} + 11\frac{5}{9}$ _____

8. $\frac{2}{4} + 6\frac{1}{12}$ _____

9. $5\frac{1}{3} + 4\frac{3}{8}$ _____

Subtract. Write the difference in simplest form.

10. $\frac{7}{11} - \frac{5}{11}$ _____

11. $2\frac{10}{13} - 1\frac{2}{3}$ _____

12. $1\frac{1}{9} - \frac{1}{3}$ _____

13. $\frac{55}{56} - \frac{7}{8}$ _____

14. $\frac{14}{15} - \frac{1}{2}$ _____

15. $2\frac{1}{5} - 2\frac{1}{10}$ _____

16. $\frac{9}{9} - \frac{1}{3}$ _____

17. $\frac{5}{9} - \frac{1}{4}$ _____

18. $\frac{5}{16} - \frac{1}{8}$ _____

Add or subtract. Write the solution in simplest form.

19. $2\frac{1}{2} + 6\frac{4}{5}$ _____

20. $\frac{1}{5} + \frac{3}{4}$ _____

21. $\frac{5}{6} - \frac{1}{8}$ _____

22. $6\frac{3}{4} - 4\frac{2}{3}$ _____

23. $1\frac{1}{6} - \frac{4}{5}$ _____

24. $3\frac{2}{5} + 5\frac{10}{10}$ _____

25. $\frac{2}{11} + \frac{7}{22}$ _____

26. $\frac{3}{5} - \frac{11}{30}$ _____

27. $7\frac{1}{6} + 9\frac{1}{2}$ _____

Basic Skills Practice

Basic Skills Practice

Adding and Subtracting Fractions, Part 2

Read each question and circle the best answer.

1. Tristan's car holds $12\frac{1}{2}$ gallons of gas. He has $2\frac{5}{6}$ gallons in it now. How many more gallons are needed to fill the tank?

   A 9 gal

   B $9\frac{1}{6}$ gal

   C $9\frac{1}{3}$ gal

   D $9\frac{2}{3}$ gal

2. Marie spends $\frac{2}{9}$ of her paycheck on rent, $\frac{1}{18}$ of her paycheck on the electric bill, and $\frac{1}{36}$ of her paycheck on the phone bill. How much of her check **remains**?

   F $\frac{11}{36}$

   G $\frac{25}{36}$

   H $\frac{7}{9}$

   I $\frac{15}{18}$

3. Anne has $1\frac{1}{2}$ ft of oak board, $2\frac{1}{3}$ ft of pine board, and $\frac{3}{4}$ ft of cedar board. How much wood does she have altogether?

   A $4\frac{7}{12}$ ft

   B $4\frac{2}{3}$ ft

   C $4\frac{3}{4}$ ft

   D 5 ft

4. Crystal and Brent have $12\frac{1}{2}$ yd of streamers for decorating a certain room. They use $5\frac{3}{4}$ yd for the back half of the room. How much is left for the front half of the room?

   F $6\frac{1}{2}$ yd

   G $6\frac{3}{4}$ yd

   H $7\frac{1}{2}$ yd

   I $7\frac{3}{4}$ yd

Solve each problem.

5. Melissa needs 4 yards of fabric for a skirt and $2\frac{3}{4}$ yards of fabric for a blouse. How many yards of fabric should Melissa buy? _____

6. A triangular-shaped garden has sides of $6\frac{1}{2}$ feet, $4\frac{2}{3}$ feet, and $5\frac{1}{3}$ feet. What is the perimeter of the garden in feet? _____

7. A shelf on a bookcase measures $3\frac{1}{3}$ feet. If books occupy $\frac{7}{8}$ feet of the bookcase, how many feet remain available for new books? _____

Basic Skills Practice

Multiplying and Dividing Fractions, Part 1

To multiply fractions, write the problem on one line. Multiply the numerators, and then multiply the denominators.

Example 1: Method 1: Multiply first, and then simplify.

$$\frac{3}{14} \times \frac{2}{3} = \frac{3 \cdot 2}{14 \cdot 3} = \frac{6}{42} = \frac{1}{7}$$

Method 2: Cancel first, and then multiply.

$$\frac{3}{14} \times \frac{2}{3} = \frac{{}^1\!\!\not{3} \cdot \not{2}^1}{{}^7\!\!\not{14} \cdot \not{3}_1} = \frac{1}{7}$$

Dividing by a fraction is the same as multiplying by the fraction's reciprocal.

Recall that the *reciprocal* of a fraction $\frac{A}{B}$ is written as $\frac{B}{A}$.

Example 2: $\frac{3}{5} \div \frac{1}{3} = \frac{3}{5} \times \frac{3}{1}$ ←Rewrite $\div \frac{1}{3}$ as $\times \frac{3}{1}$.

$= \frac{9}{5}$ ←Multiply.

$= 1\frac{4}{5}$ ←Change to a mixed number.

Remember that when you write mixed numbers as improper fractions, they follow the same rules as fractions.

Multiply. Use either method 1 or method 2. Make sure the product is in its simplest form.

1. $\frac{2}{3} \times 9$ _____

2. $\frac{4}{7} \times \frac{3}{4}$ _____

3. $\frac{1}{9} \times \frac{3}{11}$ _____

4. $\frac{2}{5} \times \frac{3}{10}$ _____

5. $\frac{4}{25} \times 50$ _____

6. $\frac{1}{15} \times \frac{5}{6}$ _____

7. $2\frac{4}{5} \times \frac{1}{8}$ _____

8. $1\frac{7}{9} \times \frac{2}{3}$ _____

9. $3\frac{1}{2} \times 1\frac{6}{7}$ _____

Divide. Write the solution in its simplest form.

10. $\frac{1}{5} \div 5$ _____

11. $\frac{3}{5} \div \frac{2}{3}$ _____

12. $\frac{7}{8} \div \frac{1}{2}$ _____

13. $\frac{2}{11} \div \frac{1}{22}$ _____

14. $2\frac{4}{9} \div 2$ _____

15. $1\frac{2}{5} \div \frac{1}{6}$ _____

16. $\frac{5}{12} \div 1\frac{3}{7}$ _____

17. $5\frac{3}{5} \div \frac{1}{5}$ _____

18. $7\frac{1}{8} \div 2\frac{5}{7}$ _____

Multiply or divide. Write the solution in its simplest form.

19. $10 \div 3\frac{1}{3}$ _____

20. $\frac{4}{5} \times \frac{10}{11}$ _____

21. $1\frac{1}{5} \times 6\frac{2}{3}$ _____

22. $\frac{4}{5} \div \frac{24}{25}$ _____

23. $2\frac{3}{4} \times 1\frac{1}{4}$ _____

24. $5\frac{1}{3} \times 2\frac{1}{4}$ _____

25. $3\frac{4}{7} \times \frac{2}{5}$ _____

26. $\frac{8}{9} \div \frac{6}{36}$ _____

27. $\frac{11}{13} \div \frac{11}{52}$ _____

Basic Skills Practice

Multiplying and Dividing Fractions, Part 2

Read each question and circle the best answer.

1. Jamal is putting a swimming pool in his backyard. The length of the pool will be $\frac{1}{3}$ of the length of the yard. If the yard is $30\frac{1}{2}$ yards, what will the length of the pool be?

   A $10\frac{1}{6}$ yards

   B $10\frac{1}{3}$ yards

   C $10\frac{1}{2}$ yards

   D $11\frac{1}{2}$ yards

2. Bonita has $1\frac{1}{2}$ pizzas to divide equally among 5 people. How much pizza will each person receive?

   F $\frac{3}{10}$ of a pizza

   G $\frac{1}{3}$ of a pizza

   H $\frac{10}{3}$ of a pizza

   I 6 pieces of pizza

3. Sally Mae baked a pie and ate a slice. There is $\frac{7}{8}$ of a pie remaining. How big of a slice did Sally eat?

   A $\frac{1}{9}$ of the pie

   B $\frac{1}{8}$ of the pie

   C $\frac{6}{8}$ of the pie

   D $\frac{7}{8}$ of the pie

4. Carlo can ride his bicycle at the rate of 15 miles per hour. How far can he go in $\frac{2}{3}$ of an hour?

   F $9\frac{2}{3}$ miles

   G $9\frac{5}{6}$ miles

   H 10 miles

   I $10\frac{1}{3}$ miles

5. It takes $1\frac{3}{4}$ cups of sugar to make a batch of cookies. How much sugar is needed to double the recipe?

   A $3\frac{1}{2}$

   B $2\frac{1}{2}$

   C $3\frac{1}{4}$

   D $2\frac{1}{4}$

6. A package of 12 hot dogs weighs $1\frac{1}{2}$ pounds. How much does each hot dog weigh?

   F $\frac{1}{8}$ of a pound

   G $\frac{1}{2}$ of a pound

   H $\frac{1}{4}$ of a pound

   I $1\frac{1}{8}$ pounds

Solve.

7. A cake recipe requires $2\frac{2}{3}$ cups of flour. If only half of the cake will fit in a pan, how many cups of flour are needed? _____

Basic Skills Practice

Cumulative Review

Read each question and circle the best answer.

1. Alex has $5\frac{1}{2}$ doggy biscuits that he wants to divide evenly among his three dogs. How many biscuits does each dog get?

   A $1\frac{4}{5}$ biscuits

   B $1\frac{5}{6}$ biscuits

   C $1\frac{1}{2}$ biscuits

   D $1\frac{2}{3}$ biscuits

2. Sylvia is a lawyer. She has just hired Frank as her legal assistant. Frank claims that he can take over $\frac{2}{5}$ of Sylvia's cases. If Sylvia estimates that she will have 325 cases for the year, how many cases would Frank handle?

   F 25 cases
   G 65 cases
   H 130 cases
   I 195 cases

3. A 19-inch TV originally priced at $240.00 is marked $\frac{1}{4}$ off. What should you multiply $240.00 by to find out how much will be saved?

   A $\frac{2}{5}$

   B $\frac{1}{4}$

   C $\frac{1}{2}$

   D $\frac{1}{5}$

4. Louie has $8\frac{1}{9}$ yd of train track, Erica has $9\frac{2}{3}$ yd, and Darrin has $7\frac{1}{3}$ yd of train track. They decide to combine the tracks to make one long track for their electric trains. How long is the combined track?

   F $24\frac{1}{9}$ yd

   G $25\frac{1}{9}$ yd

   H $25\frac{24}{27}$ yd

   I 26 yd

Solve each problem.

5. An aquarium contains $5\frac{1}{2}$ gallons of water. If Andrew adds 3 quarts of water, how much water is in the aquarium? _____

6. What is the volume of a rectangular prism if the length is 8 centimeters, the width is 2 centimeters, and the height is 4 centimeters? _____

7. What is the ***eighth*** number in the pattern: $2x^3, 4x^6, 8x^9, 16x^{12}, 32x^{15}, \ldots$? _____

Basic Skills Practice

Comparing and Ordering Numbers, Part 1

To compare numbers, start at the left and move right until the digits in the corresponding place are different.

Example 1: Compare 6744 and 6748.

674**4** _____ 674**8**   ←The numbers in the 1000s, 100s, and 10s places are the same.

4 _____ 8   ←Compare the numbers in the ones place.

4 < 8   ←Therefore 6744 < 6748.

Example 2: Compare 0.457 and 0.45.

0.45**7** _____ 0.45**0**   ←Add a zero to 0.45 to show 3 places to the right of the decimal.

7 _____ 0   ←Compare the numbers in the thousandths place.

7 > 0   ←0.457 > 0.45

To arrange 3 numbers in order, compare twice.

Example 3: Arrange 867.18, 865.53, and 867.05 in order from greatest to least.

86**7**.18 _____ 86**5**.53 _____ 86**7**.05   Since 5 < 7, 865.53 is the least.

867.**1**8 _____ 867.**0**5   Since 1 > 0, 867.18 is the greatest.

The numbers in order from greatest to least are 867.18, 867.05, and 865.53.

Compare. Write =, <, or >.

1. 852 _____ 853

2. 5000 _____ 4999

3. 8459 _____ 8495

4. 76,217 _____ 76,712

5. 0.04 _____ 0.4

6. 0.09 _____ 0.009

7. 7.5 _____ 0.75

8. 71.4 _____ 71.5

9. 518.567 _____ 518.548

10. 11.2579 _____ 10.2579

11. 5.2786 _____ 5.27865

12. 40,847.6 _____ 40,847.68

Arrange in order from greatest to least.

13. 8642; 8462; 8536

14. 0.0896; 0.3469; 0.1452

_____

_____

Arrange in order from least to greatest.

15. 48,492; 486,743; 48,619

16. 0.489; 4.89; 0.49

_____

_____

17. Sonia is 5 feet 8 inches tall, Mary is 5 feet 3 inches, Sue is 5 feet 2 inches, and Tony is 6 feet 2 inches. List the names from tallest to shortest.

_____, _____, _____, _____

Basic Skills Practice

Comparing and Ordering Numbers, Part 2

Read each question and circle the best answer.

1. Juan's job at the school library is to shelve the books. If the book numbers are arranged from smallest to largest, which shows the correct order?

   A 518.2, 518.35, 518.885, 518.8
   B 518.35, 518.885, 518.8, 518.2
   C 518.8, 518.35, 518.2, 518.885
   D 518.2, 518.35, 518.8, 518.885

2. To prepare a graph, students were instructed to order their data from smallest to largest. Which list shows the correct order?

   F 70, 60.2, 70.8, 70.09
   G 60.2, 70, 70.09, 70.8
   H 70.09, 70.8, 70, 60.2
   I 60.2, 70.09, 70.8, 70

3. Four of the largest cities in Texas have populations of 935,933; 1,630,553; 515,342; and 1,006,877. Which list shows the populations in order from largest to smallest?

   A 515,342; 1,630,553; 935,933; and 1,006,877
   B 935,933; 1,630,553; 515,342; and 1,006,877
   C 1,630,553; 1,006,877; 935,933; and 515,342
   D 1,006,877; 1,630,553; 935,933; and 515,342

4. Which list shows times in order from least to greatest?

   F 45.72 s; 45.50 s; 45.59 s; 59.63 s
   G 45.50 s; 45.59 s; 45.72 s; 59.63 s
   H 59.63 s; 45.50 s; 45.72 s; 45.59 s
   I 59.63 s; 45.72 s; 45.59 s; 45.50 s

5. The average yearly rainfall for four cities is 35.74 in., 36.59 in., 36.97 in., and 36.53 in. Which list shows the rainfall amounts in order from greatest to least?

   A 35.74, 36.59, 36.53, 36.97
   B 36.97, 36.53, 36.59, 35.74
   C 36.59, 36.53, 36.97, 35.74
   D 36.97, 36.59, 36.53, 35.74

6. Four students lined up from shortest to tallest. Which list shows that order?

   F 1.4775 m, 1.4725 m, 1.475 m, 1.4675 m
   G 1.4675 m, 1.4725 m, 1.475 m, 1.4775 m
   H 1.4675 m, 1.475 m, 1.4775 m, 1.4725 m
   I 1.4725 m, 1.475 m, 1.4675 m, 1.4775 m

7. While shopping for school supplies, Maria saw four different types of notebooks. They ranged in price from $1.29 to $3.65. Which list below shows the prices of the notebooks from least to greatest?

   A $1.29, $1.95, $3.65, $2.59
   B $1.95, $2.59, $3.65, $1.29
   C $3.65, $2.59, $1.95, $1.29
   D $1.29, $1.95, $2.59, $3.65

8. Scores for the all-around event at a high school gymnastics meet were 39.711, 39.848, 39.679, and 39.736. Which list shows these scores in order from greatest to least?

   F 39.848, 39.736, 39.711, 39.679
   G 39.848, 39.711, 39.736, 39.679
   H 39.679, 39.711, 39.736, 39.848
   I 39.679, 39.848, 39.736, 39.711

Basic Skills Practice

Adding and Subtracting Decimals, Part 1

When you add or subtract decimals, first align the decimal points.

Example 1: Add: 4.36 + 7.8 + 46.837.

First, align the numbers, using the decimal points as a guide:

$$\begin{array}{r} 4.36 \\ 7.8 \\ + 46.837 \end{array}$$

Second, use zeros as placeholders:
Remember: 4.36 = 4.360
7.8 = 7.800
Then add:

$$\begin{array}{r} 4.36\mathbf{0} \\ 7.8\mathbf{00} \\ + 46.837 \\ \hline 58.997 \end{array}$$

Example 2: Subtract: 56 − 7.58.

First, align the numbers, using the decimal points as a guide:

$$\begin{array}{r} 56. \\ - 7.58 \end{array}$$

Second, use zeros as placeholders:
Remember: 56 = 56.00
Then subtract:

$$\begin{array}{r} 56.\mathbf{00} \\ - 7.58 \\ \hline 48.42 \end{array}$$

Add.

1. 6.8 + 1.8 _____

2. 9.18 + 6.45 _____

3. 4.007 + 0.372 _____

4. 17.36 + 2.44 _____

5. 1.507 + 66.005 _____

6. 0.9 + 15.67 _____

7. 2.863 + 26.24 _____

8. 9.36 + 54.29 + 4.7 _____

9. 5.56 + 2.39 + 24.8 _____

10. 25.6 + 4.8 + 35.07 _____

Subtract.

11. 0.7 − 0.5 _____

12. 8.87 − 6.53 _____

13. 4.28 − 0.5 _____

14. 0.547 − 0.45 _____

15. 7.594 − 6.98 _____

16. 7.85 − 4.99 _____

17. 8.204 − 4.865 _____

18. 5.2 − 0.87 _____

19. 4.5 − 2.57 _____

20. 17.5 − 5.681 _____

Solve.

21. During the winter months in a northeastern city, 5.4 in., 4.7 in., and 10.2 in. of snow fell. What was the total amount of snowfall? _____

Basic Skills Practice

Adding and Subtracting Decimals, Part 2

Read each question and circle the best answer.

1. David ordered a belt for $15.27, a pair of pants for $24.30, and a hat for $17.25 from a catalog. He added $11.85 for tax, shipping, and handling. What was the total cost of David's order?

   A $68.67

   B $68.57

   C $67.67

   D $58.67

2. The Juarezes drove their family car on vacation. The mileage on the odometer read 189.7 when they started. When they returned, the reading was 875.9. How many miles did the Juarezes drive on their vacation?

   F 684.2 mi

   G 686.2 mi

   H 696.2 mi

   I 714.2 mi

3. On a four day trip, Ahmed drove 269.4 miles the first day, 354.2 miles the second day, 283 miles the third day, and 185.8 miles the fourth day. What was the total distance he traveled?

   A 837.7 mi

   B 992.4 mi

   C 1082.4 mi

   D 1092.4 mi

4. Eric bought a pair of jeans that were on sale for $8.75 off the original price. At the register, the jeans were reduced another $2.50. If the original price was $34.50, how much did Eric pay for the jeans?

   F $ 23.25

   G $ 24.50

   H $ 25.75

   I $ 32.00

5. At Pet Warehouse, Marco bought a flea collar for $5.39 and 2 bags of dog food for $11.98 each. If he paid $2.35 in sales tax, how much did he spend altogether?

   A $19.72

   B $21.70

   C $31.70

   D $59.72

6. Natalie had $137.54 in her checking account. After writing a check for $45.68, how much money did she have left?

   F $81.86

   G $91.76

   H $91.86

   I $92.14

Solve each problem.

7. Lisa purchased a sweater for $38.95 and 2 bracelets for $6.95 each. The sales tax on the purchase was $3.45. What was the total amount she spent?   _____

8. A can of dried fruit contains 59.3% apples, 16.1% apricots, 10.4% raisins, 8.4% dates, and 5.8% figs. What percent of the can is apricots, raisins, and figs?   _____

Basic Skills Practice

Cumulative Review

Read each question and circle the best answer.

1. Which is equivalent to $3^2 < 4^3$?

   A $3 \times 2 < 4 \times 3$

   B $2 \times 3 < 3 \times 3$

   C $6 < 12$

   D $3 \times 3 < 4 \times 4 \times 4$

2. Which group of fractions is in order from least to greatest?

   F $\dfrac{1}{4}, \dfrac{2}{3}, \dfrac{5}{6}, \dfrac{5}{9}$

   G $\dfrac{2}{3}, \dfrac{1}{4}, \dfrac{5}{6}, \dfrac{5}{9}$

   H $\dfrac{5}{9}, \dfrac{2}{3}, \dfrac{5}{6}, \dfrac{1}{4}$

   I $\dfrac{1}{4}, \dfrac{5}{9}, \dfrac{2}{3}, \dfrac{5}{6}$

3. From 2 gallons of oil, Juan used 5 quarts of oil for his car. How many **quarts** are left?

   A 2 qt

   B 3 qt

   C 4 qt

   D 5 qt

4. What expression should come next in the pattern: $3w, 6w, 9w, 12w,$ _____ ?

   F $18w$

   G $15w$

   H $12w$

   I $10w$

5. In a long-jump competition, the distances, in feet, are posted in order from greatest to least. Which list shows the correct order?

   A 77.4, 76.38, 77.42, 76.374

   B 76.374, 76.38, 77.42, 77.4

   C 77.42, 77.4, 76.38, 76.374

   D 76.38, 76.374, 77.4, 77.42

6. Susan bought $13.37 worth of groceries. She paid for it with a $20 bill. How much change did she receive?

   F $33.37

   G $7.63

   H $6.63

   I $6.37

7. A store display was created using boxes of different sizes placed in order from shortest to tallest. Which list shows that order?

   A 0.76 m, 0.84 m, 0.48 m, 0.67 m

   B 0.84 m, 0.76 m, 0.67 m, 0.48 m

   C 0.48 m, 0.67 m, 0.76 m, 0.84 m

   D 0.67 m, 0.48 m, 0.84 m, 0.76 m

8. Kate drove 295 miles in 5.1 hours. Estimate her average speed.

   F 40 mph

   G 50 mph

   H 60 mph

   I 70 mph

Solve each problem.

9. Dante stuffs 25 envelopes every 4 minutes. How many can he stuff in one hour? _____

10. The area of 4 tablemats is $A = 4lw$, where l is the length and w is the width. When l is 20 in. and w is 14 in., what is the total area for the 4 tablemats? _____

Basic Skills Practice

Multiplying and Dividing Decimals, Part 1

Example 1: Multiply 0.57 × 0.075.
First put the number with the most digits on top. Then multiply as for whole numbers.

$$\begin{array}{r} 0.075 \\ \times\ .57 \\ \hline 00525 \\ 00375\ \\ \hline 004275 \end{array}$$

Second count the decimal places in the factors.

$$\begin{array}{r} 0.075 \leftarrow 3\text{ decimal places} \\ \times\ 0.57 \leftarrow 2\text{ decimal places} \\ \hline 0.04275 \leftarrow 5\text{ decimal places} \end{array}$$

↑ Insert zeros as place holders if needed.

Example 2: Divide 28.8 ÷ 0.8.
First move the decimal point as needed to make the divisor a whole number. Move the decimal point in the dividend the same number of places.

$$0.8\overline{)28.8}$$

Second divide. Align the decimal point of the quotient with that of the dividend.

$$8.\overline{)288.}\quad 36.$$

$$\begin{array}{r} 24\ \\ \hline 48 \\ 48 \\ \hline 0 \end{array}$$

Multiply.

1. 0.7 × 8 _____

2. 0.8 × 6 _____

3. 0.5 × 9 _____

4. 0.7 × 0.9 _____

5. 3.9 × 5 _____

6. 2.5 × 8.5 _____

7. 4.7 × 6.5 _____

8. 5.6 × 7.8 _____

9. 8.6 × 7.48 _____

10. 4.8 × 7.27 _____

Divide.

11. 6.3 ÷ 9 _____

12. 5.2 ÷ 5 _____

13. 719.2 ÷ 29 _____

14. 103.6 ÷ 0.4 _____

15. 352.8 ÷ 0.7 _____

16. 106.83 ÷ 0.09 _____

17. 0.00578 ÷ 0.02 _____

18. 38.8275 ÷ 0.465 _____

19. 18.928 ÷ 0.52 _____

20. 262.08 ÷ 2.52 _____

Basic Skills Practice

Multiplying and Dividing Decimals, Part 2

Read each question and circle the best answer.

1. Mark bought 75 feet of rope for $18. How much would 1 foot of rope cost?

   A $0.024
   B $0.24
   C $0.26
   D $0.42

2. A plane flew 500 miles in 40 minutes. How long would it take to fly 1 mile?

   F 12.5 min
   G 8 min
   H 1.25 min
   I 0.08 min

3. A 32-ounce bottle of ketchup sells for $1.28. What is the price per ounce?

   A $0.40
   B $0.15
   C $0.04
   D $0.02

4. A real estate agent charges 5% commission on the sale price of a house. How much is the commission on the sale of a $150,000 house?

   F $550
   G $750
   H $5500
   I $7500

5. One gallon of orange juice weighs 6.18 pounds. How much would a dozen gallons of orange juice weigh?

   A 18.54 lb
   B 61.8 lb
   C 64.16 lb
   D 74.16 lb

6. Lauren put 15 gallons of gasoline in her car at a cost of $1.35 per gallon. How much did Lauren spend for gasoline?

   F $20.25
   G $20.00
   H $19.25
   I $18.10

7. Jerry paid $16.77 for 13 gallons of gasoline. What was the cost per gallon?

   A $1.17
   B $1.19
   C $1.27
   D $1.29

8. Franklin is buying pizzas for a class party. Each pizza costs $11.25, including tax, and feeds 5 students. How much money will he need to feed 20 students?

   F $90
   G $75
   H $60
   I $45

Solve each problem.

9. The price of a stereo was $820. The price has been marked down 33 percent. How much money will be saved at the sale price? _____

10. Janice bought some binders at $1.97 each. Her total was $17.73, before tax. How many binders did she buy? _____

Basic Skills Practice

Rounding Whole Numbers and Decimals, Part 1

To round a number to a given place, look at the digit to the right of that place. If it is less than 5, the digit stays the same. If it is 5 or greater, the digit is changed to the next higher digit. Rounding to the ones place is often referred to as rounding to the nearest whole number.

Example 1: Round 47,678 to the nearest thousand.
4<u>7</u>,678   ← Underline the number in the thousands place.
4<u>7</u>,**6**78   ← Look at the digit to the right of the underlined digit; 6 is greater than 5.
4<u>8</u>,000   ← Round up by increasing the digit in the thousands place by 1 and changing all the digits to the right to zeros.

When rounding decimals, follow the same steps and drop the digits to the right of the given place.

Example 2: Round 5.942 to the nearest hundredth.
5.9<u>4</u>2   ← Underline the number in the hundredths place.
5.9<u>4</u>**2**   ← Look at the digit to the right; 2 is less than 5.
5.9<u>4</u>   ← Keep the digit in the hundredths place the same. *Drop* the digits to the right of the hundredths place.

Round to the nearest hundred.

1. 4596 _____

2. 8327 _____

3. 15,209 _____

4. 96,785 _____

Round to the nearest thousand.

5. 4498 _____

6. 22,845 _____

7. 88,397 _____

8. 697,573 _____

Round to the nearest one or whole number.

9. 45.6 _____

10. 9.45 _____

11. 76.854 _____

12. 23.59 _____

Round to the nearest tenth.

13. 5.872 _____

14. 37.345 _____

15. 486.74 _____

16. 65.39 _____

Round to the nearest hundredth.

17. 6.543 _____

18. 45.386 _____

19. 0.767 _____

20. 16.452 _____

Basic Skills Practice

Rounding Whole Numbers and Decimals, Part 2

Read each question and circle the best answer.

1. In an encyclopedia, the population of a country is rounded to the nearest thousand. If Canada's population is rounded to 26,832,000, which of the following could **not** be Canada's actual population?

   A 26,832,572

   B 26,832,366

   C 26,832,299

   D 26,831,824

2. What is 397.3 rounded to the nearest whole number?

   F 390

   G 397

   H 398

   I 400

3. In an almanac, the population of small towns is rounded to the nearest hundred. How would the population of 28,764 be given?

   A 30,000

   B 28,800

   C 28,760

   D 28,700

4. A company's stock was priced at $14.875 at the end of today's business. What is this amount rounded to the nearest cent?

   F $14.75

   G $14.87

   H $14.88

   I $14.90

5. A baseball player's batting average is calculated to be 0.36784. What is this average rounded to the nearest thousandth?

   A 0.3678

   B 0.368

   C 0.367

   D 0.366

6. During a math lesson, students were asked to round different measurements to the nearest whole number. How should 66.4 cm be rounded?

   F 60 cm

   G 66 cm

   H 67 cm

   I 70 cm

7. Catherine ran two miles in 23.54 minutes. What is this time rounded to the nearest tenth?

   A 24.6 min

   B 24.0 min

   C 23.6 min

   D 23.5 min

8. A pilot flew his aircraft at 85,068.997 feet. What is this amount rounded to the nearest whole number?

   F 85,070 ft

   G 85,069 ft

   H 85,068 ft

   I 85,060 ft

Solve.

9. A major computer company announced that it earned $45,894,352 last year. What is this amount rounded to the nearest thousand? _____

Basic Skills Practice
Cumulative Review

Read each question and circle the best answer.

1. What is the sum of 0.347 + 1.866 rounded to the nearest tenth?

  A 2.0

  B 2.2

  C 2.21

  D 2.213

2. A recipe calls for $6\frac{1}{2}$ cups of flour. The baker has a $\frac{1}{2}$-cup measure in the kitchen. How many times should she fill the measuring cup to obtain the needed amount of flour?

  F $6\frac{1}{2}$

  G 7

  H 12

  I 13

3. Each of 6 friends pays equal amounts of a restaurant bill that is $51.72, including tip. How much does each friend pay?

  A $8.02

  B $8.60

  C $8.62

  D $9.02

4. The senior class needs 325 feet of ribbon to decorate the gymnasium for a dance. To allow for mistakes, the chairperson bought 125 yards of ribbon. How many extra *feet* of ribbon did the chairperson purchase?

  F 25 ft

  G 50 ft

  H 200 ft

  I 450 ft

5. An equilateral triangle has a perimeter of 53.1 centimeters. What is the length of one of its sides?

  A 159.3 cm

  B 18.7 cm

  C 17.7 cm

  D 17.6 cm

6. Vince went to Burger Delite for dinner. He ordered a Jumbo Burger for $2.19, medium fries for $0.89, and a vanilla shake for $1.19. Tax on his order was $0.35. What was the total cost of his order?

  F $3.94

  G $4.27

  H $4.62

  I $5.84

7. An art teacher is cutting a piece of string 48 feet long into equal pieces to give to 15 students. How long should each piece be?

  A 4.2 ft

  B 3.3 ft

  C 3.25 ft

  D 3.2 ft

8. On a quiz show, prize values are rounded to the nearest dollar. How will a prize valued at $846.51 be rounded?

  F $900

  G $847

  H $846

  I $800

Basic Skills Practice

Decimals and Percents, Part 1

You change a decimal to a percent by multiplying the decimal by 100 and adding a percent sign.

Example 1: Multiplying by 100 moves the decimal point two places to the right.
$0.34 \times 100 = 34$, so $0.34 = 34\%$

You change a percent to a decimal by dividing the percent by 100 and omitting the percent sign.

Example 2: Dividing by 100 moves the decimal point two places to the left.
$25 \div 100 = 0.25$, so $25\% = 0.25$

Write each decimal as a percent.

1. 0.05 _____

2. 0.22 _____

3. 0.75 _____

4. 0.17 _____

5. 0.41 _____

6. 0.03 _____

Write each percent as a decimal.

7. 82% _____

8. 37% _____

9. 7% _____

10. 55% _____

11. 3% _____

12. 68% _____

Write a percent for each decimal.

13. 0.544 _____

14. 0.0065 _____

15. 0.085 _____

16. 0.105 _____

17. 0.014 _____

18. 1.75 _____

Write a decimal for each percent.

19. 3.4% _____

20. 17.8% _____

21. 100% _____

22. 83% _____

23. 0.16% _____

24. 755.2% _____

Solve each problem.

25. What is 5.7% written as a decimal? _____

26. What is the correct way to write 0.75 as a percent? _____

Basic Skills Practice

Decimals and Percents, Part 2

Read each question and circle the best answer.

1. 0.50 written as a percent is—

   A 0.50%

   B 5%

   C 50%

   D 500%

2. The percent for 0.162 is—

   F 162%

   G 16.2%

   H 1.62%

   I 0.162%

3. Approximately 0.21 of the Earth's atmosphere is oxygen. What percent of the Earth's atmosphere is oxygen?

   A 210%

   B 21%

   C 2.1%

   D 0.21%

4. The decimal for 103% is—

   F 0.103

   G 1.03

   H 10.3

   I 10.5

5. A company estimates that its profit for a year is 20% of its sales for that year. What is 20% written as a decimal?

   A 0.12

   B 0.2

   C 0.02

   D 0.002

6. Of the students interviewed, 55% prefer sausage pizza. What is this percent written as a decimal?

   F 55.0

   G 5.5

   H 0.55

   I 0.50

7. About 0.03 of the Earth's water is fresh water. What percent is fresh water?

   A 30%

   B 3%

   C 0.3%

   D $\frac{1}{3}$%

Solve each problem.

8. Shireen's family is buying a more efficient air conditioner. It will use 23.5% less energy. What is this percent written as a decimal? _____

9. A new word-processing program is supposed to be more powerful and increase productivity. It is more powerful by a factor of 0.365. What is this written as a percent? _____

10. The Channel 14 weather forecast announced that the probability of rain tomorrow is 60%. What is the correct way to express this percentage as a decimal? _____

Basic Skills Practice

Writing Percents as Fractions, Part 1

A percent is a ratio that compares a number with 100. To write a percent as a fraction, write the number as the numerator and 100 as the denominator.

41% means 41 out of 100. $41\% = \frac{41}{100}$

7% means 7 out of 100. $7\% = \frac{7}{100}$

After you write the fraction, reduce it to lowest terms.

25% means 25 out of 100. $25\% = \frac{25}{100} = \frac{1}{4}$

Percents may be greater than 100. Write as an improper fraction, and then reduce.

110% means $110 \div 100$. $110\% = \frac{110}{100} = \frac{11}{10} = 1\frac{1}{10}$

Write each percent as a fraction.

1. 13% _____
2. 23% _____
3. 37% _____

4. 11% _____
5. 3% _____
6. 99% _____

7. 49% _____
8. 7% _____
9. 39% _____

Write each percent as a fraction. Then reduce to lowest terms.

10. 50% _____
11. 75% _____

12. 66% _____
13. 12% _____

14. 24% _____
15. 40% _____

Write each percent as a fraction or mixed number in reduced form.

16. 160% _____
17. 175% _____

18. 120% _____
19. 100% _____

20. 113% _____
21. 500% _____

Solve each problem.

22. Kay has served as mayor for 75% of her term. What fraction of her term has she served? _____

23. Mr. Renwiz phoned 27 students to tell them that the school was closed due to flooding. Each call took approximately 10% of an hour. What fraction of an hour was Mr. Renwiz on the phone for each call? _____

Basic Skills Practice

Writing Percents as Fractions, Part 2

Read each question and circle the best answer.

1. 83% written as a fraction is—

    A $\frac{8.3}{100}$

    B $\frac{83}{100}$

    C $\frac{83}{83}$

    D $\frac{83}{10}$

2. Which fraction is equal to 25%?

    F $\frac{2}{15}$

    G $\frac{1}{4}$

    H $\frac{1}{3}$

    I $\frac{1}{2}$

3. 43 of 100 people surveyed eat in restaurants once every weekend. What percent is this?

    A 57%

    B 43%

    C 4.3%

    D 0.43%

4. Ten percent of the class made an A. This percent is equal to—

    F $\frac{1}{3}$

    G $\frac{1}{4}$

    H $\frac{1}{5}$

    I $\frac{1}{10}$

5. A 50% discount is equal to—

    A $\frac{3}{8}$

    B $\frac{1}{2}$

    C $\frac{2}{3}$

    D $\frac{4}{5}$

6. 100 surveys were sent out randomly; 45% of them were returned. How many surveys were returned?

    F 100

    G 54

    H 53

    I 45

Solve each problem.

7. On a field trip, 50% of the students bought something to eat before boarding the bus. What fraction of the students bought something to eat? _____

8. Arthur made 75% of 20 attempted free throws last season. What fraction of the shots he attempted did he make? _____

9. In the Student Government election, only 55% of the students voted. What fraction of the students voted? _____

Basic Skills Practice
Cumulative Review

Read each question and circle the best answer.

1. The decimal 0.52 written as a percent is—

   A $\frac{1}{2}$%

   B 0.52%

   C 5.2%

   D 52%

2. Tara bought two sweaters. One sweater cost $19.95. The second sweater cost $3.55 more. How much did the second sweater cost?

   F $16.40

   G $22.40

   H $23.40

   I $23.50

3. Evaluate $-x + 9$ for $x = 4$.

   A 13

   B 5

   C −5

   D −13

4. The percent equivalent to $\frac{21}{100}$ is—

   F 21%

   G 20%

   H 0.21%

   I 0.021%

5. 0.558 rounded to the nearest tenth is—

   A 0.6

   B 0.55

   C 0.56

   D 0.5

6. A recipe calls for $1\frac{1}{4}$ cup of sugar and $1\frac{1}{3}$ cup butter. Which is greater, the amount of sugar or the amount of butter?

   F flour

   G butter

   H sugar

   I They are equal.

7. Percent is a ratio comparing a number with—

   A itself

   B 100

   C sales tax

   D a fraction

8. Yoshi's car travels 18 miles on a gallon of gas. How far could the car travel on 17.6 gallons of gas?

   F 35.6 mi

   G 316.8 mi

   H 356 mi

   I 3168 mi

Solve each problem.

9. Which is the largest number: 0.08 or 0.8? _____

10. What is the number 1.567 rounded to the nearest tenth? _____

11. Mary needs to cut $3\frac{1}{3}$ feet from a $5\frac{1}{3}$ foot board. How much board is left? _____

Basic Skills Practice

Fractions, Decimals, and Percents, Part 1

A percent is another way to represent a fraction or a decimal.

Example 1: 25% of the student population participates in an after school sports program.
$$25\% = \frac{25}{100} = \frac{1}{4} \qquad 25\% = 25 \div 100 = 0.25$$

Example 2: Angie spent $\frac{1}{8}$ of her budget on movie tickets.

To change $\frac{1}{8}$ to a percent, first divide 1 by 8 to make an equivalent decimal, 0.125.

Then multiply by 100 and add a percent sign.
$$\frac{1}{8} = 0.125 = 12.5\%$$

Write a fraction and a decimal for each percent.

1. 45% _____

2. 75% _____

3. 5% _____

4. 32% _____

5. 80% _____

6. 12% _____

Write a percent for each fraction or decimal.

7. $\frac{79}{100}$ _____

8. 0.82 _____

9. 0.07 _____

10. $\frac{26}{100}$ _____

11. $\frac{168}{100}$ _____

12. 0.11 _____

Write a percent for each fraction.

13. $\frac{1}{5}$ _____

14. $\frac{3}{10}$ _____

15. $\frac{1}{2}$ _____

16. $\frac{4}{25}$ _____

17. $\frac{9}{20}$ _____

18. $\frac{17}{50}$ _____

Solve each problem.

19. Ellen plans to finance her new car. The dealer offers an annual interest rate of 9.9%. The bank offers an annual interest rate of $9\frac{3}{4}$%. Which is the lower interest rate?

_____

20. Juan won 26 of 40 games in a tennis tournament. What percent of the games did he win?

_____

Basic Skills Practice

Fractions, Decimals, and Percents, Part 2

Read each question and circle the best answer.

1. During a recent game, Gene scored $\frac{1}{10}$ of the total points scored. What percent of the points did Gene score?

   A 10%

   B 25%

   C 50%

   D 90%

2. During the school day, the computer lab is being used 95% of the time. What part of the school day is the lab being used?

   F $\frac{1}{2}$

   G $\frac{2}{3}$

   H $\frac{3}{4}$

   I $\frac{19}{20}$

3. Arthur spends $0.57 of every dollar he earns on his baseball card collection. What percent of his money does Arthur spend on baseball cards?

   A 57%

   B 50%

   C 5.7%

   D 0.57%

4. Three-fourths of the swim team is over the age of 16. What percent of the team is over 16?

   F 75%

   G 50%

   H 34%

   I 10%

5. Melissa answered 89 of 100 questions on the science test correctly. What percent of the questions did Melissa answer correctly?

   A 11%

   B 50%

   C 89%

   D 98%

6. Misha answered 15% of 100 questions on the science test incorrectly. What fraction represents the questions that Misha answered incorrectly?

   F $\frac{17}{20}$

   G $\frac{3}{4}$

   H $\frac{3}{5}$

   I $\frac{3}{20}$

Solve each problem.

7. A recent survey showed that $\frac{1}{2}$ of the students surveyed would like school to begin after Labor Day. Show this fraction as a decimal. _____

8. Rashel's grandmother recently celebrated her 100th birthday. On her birthday cake, 87 of the candles were pink and the remainder were other colors. What percent of the candles were other colors? _____

9. Mrs. Lee received a $250 bonus with her monthly paycheck. Her regular monthly check is $2500. What percent of her regular monthly pay is her bonus? _____

Basic Skills Practice

Finding a Percent of a Number, Part 1

To find a percent of a number, change the percent to a decimal and multiply.

Example 1: A catcher's mitt is on sale for 33% off the original price of $45.00. How much money will be saved?

$33\% = 0.33$ ←Change the percent to a decimal.

$0.33 \times 45 = 14.85$ ←Multiply the decimal by the original amount.

33% of $45 is $14.85. This is the amount of savings.

Example 2: You buy a CD for $13.99. The sales tax is 8%. How much is the sales tax?

$8\% = 0.08$ ←Change the percent to a decimal.

$0.08 \times 13.99 = 1.1192$ ←Multiply the decimal by the amount of the item.

8% of $13.99 is $1.12 because you must round the number up. The sales tax is $1.12.

Compute.

1. What is 20% of 35? _____
2. What is 35% of 60? _____
3. What is 10% of 90? _____
4. What is 200% of 17? _____
5. What is 50% of 186? _____
6. What is 60% of 95? _____
7. What is 5% of 800? _____
8. What is 12% of 7300? _____
9. What is 33% of 142? _____
10. What is 67% of 15? _____
11. Find 6% of 895. _____
12. Find 32% of 96. _____
13. Find 25% of 260. _____
14. Find 75% of 192. _____
15. Find 150% of 16. _____
16. Find 60% of 90. _____
17. Find 40% of 80. _____
18. Find 55% of 20. _____
19. Find 3% of 300. _____
20. Find 12.5% of 72. _____

Solve each problem.

21. A farmer expects to increase his crop yield by 38%. If he harvested 285 bushels of corn last year, how many more bushels should he expect this year? _____

22. Beth sold a color enlargement in a wooden frame for $150. Sales tax is 8%. How much was the sales tax? _____

23. Lu makes $8.75 an hour. He is getting a 20% raise. How much will his raise be per hour? _____

Basic Skills Practice

Finding a Percent of a Number, Part 2

Read each question and circle the best answer.

1. Approximately 67% of your body weight is water. If your weight were 125 pounds, how many pounds would be water?

    A 58.0 lbs

    B 67.5 lbs

    C 83.75 lbs

    D 87.36 lbs

2. In the United States, about 46% of the population wears contact lenses or glasses. About how many students in a school of 1200 would you expect to wear glasses or contact lenses, based on this data?

    F 425

    G 500

    H 552

    I 1000

3. Angela spends 25% of her day in school. How many hours does Angela spend in school each day? Use a 24-hour day.

    A 9 h

    B 8 h

    C 7 h

    D 6 h

4. The baseball team that is currently in first place has won 75% of their 28 games this year. How many games have they won?

    F 20

    G 21

    H 25

    I 26

5. About 40% of children and teenagers bite their nails. A school district has an enrollment of 16,880 students. Based on this data, about how many students are probably nail-biters?

    A 10,128

    B 1012.8

    C 6752

    D 675.2

6. The weatherman reports that there is a 12% decrease in the average temperature from summer to fall. The average temperature during the summer was 95 degrees. By how much will the average temperature decrease in the fall?

    F 12.0°

    G 11.4°

    H 1.20°

    I 1.14°

Solve each problem.

7. Mr. Hamilton is planning a graduation party for 60 soccer players. He will be serving soft drinks and cookies. If 85% of the players have said they will be coming to the party, how many soft drinks should Mr. Hamilton order? _____

8. If each player coming to Mr. Hamilton's party brings a guest, how many soft drinks should Mr. Hamilton order? _____

Basic Skills Practice

Cumulative Review

Read each question and circle the best answer.

1. $\frac{6}{10}$ written as a percent is—

   A 60%

   B 68%

   C 75%

   D 78%

2. 58% of 72 is—

   F 31.76

   G 41.76

   H 41.82

   I 58.72

3. $18 − $9.63 =

   A $9.47

   B $9.37

   C $8.37

   D $7.47

4. 0.208, 0.028, 0.28, and 0.2 ordered from least to greatest are—

   F 0.28, 0.208, 0.028, and 0.2

   G 0.208, 0.028, 0.28, and 0.2

   H 0.28, 0.208, 0.2, and 0.028

   I 0.028, 0.2, 0.208, and 0.28

The table below shows the responses of 1000 students who were asked whether they spend too much or too little time reading for pleasure.

Too much	Too little	About right	Don't know
7%	73%	16%	4%

5. The number of students who said they read too much is—

   A 730

   B 70

   C 16

   D 7

6. The number of students who responded "don't know" is—

   F 730

   G 160

   H 40

   I 4

Solve each problem.

7. A computer diskette has a mass of 2.5 grams. How many milligrams is the mass of the disk? _____

8. Simplify the expression: $0 − 3(5 − 4) − 6$. _____

9. What is the product of 0.048×0.9? _____

10. Most credit cards charge 1.5% interest per month on the unpaid balance. If the balance is $300, what is the interest charge for one month? _____

Basic Skills Practice

Finding the Percent One Number Is of Another, Part 1

To find what percent one number is of another, write a number sentence and solve.
To convert the decimal to a percent, multiply by 100.

Example: What percent of 50 is 25?

$$n \times 50 = 25 \quad \text{Write a number sentence.}$$

$$\frac{n \times 50}{50} = \frac{25}{50} \quad \text{Solve.}$$

$$n = 0.5 = 50\% \quad \text{Rename.}$$

Write a number sentence and solve to find the percent.

1. What percent of 55 is 11?

   Solve: _____

   Number sentence: _____

   Rename: $n =$ _____ = _____%

2. What percent of 24 is 6?

   Solve: _____

   Number sentence: _____

   Rename: $n =$ _____ = _____%

3. What percent of 25 is 20?

   Solve: _____

   Number sentence: _____

   Rename: $n =$ _____ = _____%

4. What percent of 40 is 26?

   $n =$ _____

5. What percent of 120 is 42?

   $n =$ _____

Solve each problem.

6. $n\%$ of 25 is 12; $n =$ _____

7. $n\%$ of 96 is 48; $n =$ _____

8. $n\%$ of 84 is 21; $n =$ _____

9. $n\%$ of 48 is 36; $n =$ _____

10. Thirty students ordered pizza. Six of them like mushrooms on their
    pizza. What percent of the students like mushrooms on their pizza? _____

11. On the last test, Rachel answered 20 out of 25 questions correctly.
    What percent of problems did Rachel answer ***incorrectly***? _____

12. Michael made 3 out of 15 free throws. What percent of the free
    throws did he ***miss***? _____

Basic Skills Practice

Finding the Percent One Number Is of Another, Part 2

Read each question and circle the best answer.

1. What percent of 36 is 27?

   A 7%

   B 7.5%

   C 70%

   D 75%

2. What percent of 40 is 24?

   F 6%

   G 50%

   H 60%

   I 66%

3. Darius ate 15 out of 20 grapes. What percent of the grapes did he eat?

   A 7.5%

   B 70%

   C 75%

   D 80%

4. n% of 45 is 27; n is—

   F 45%

   G 50%

   H 60%

   I 66%

5. Angie answered 32 of 40 questions correctly. What percent of the answers did she get correct?

   A 80%

   B 85%

   C 88%

   D 90%

6. What percent of 80 is 20?

   F 2.5%

   G 4%

   H 25%

   I 40%

Solve each problem.

7. Gene went to the mall with $60. He spent $12 playing games. What percent did Gene spend? _____

8. What percent of 75 is 30? _____

9. Four out of five people watch television for more than one hour each day. What percent of 5 is 4? _____

10. There are 45 cookies on a plate; 36 of the cookies are chocolate chip. What percent of the cookies are chocolate chip? _____

11. Jamal is saving his money to buy his mother a new bracelet. He has saved $90 of the $150 he needs. What percent has he saved? _____

12. This morning the bagel shop sold 180 of the bagel of the day; 99 of the bagels were sold with cream cheese. What percent of the bagel of the day were sold with cream cheese? _____

Basic Skills Practice

Finding a Number When a Percent of It Is Known, Part 1

To find a number when a percent of it is known, write a number sentence that renames the percent as a decimal and solve.

Example: 75% of what number is 150?

$$\downarrow \quad \downarrow \qquad \downarrow \qquad \qquad \downarrow \quad \downarrow$$

$$0.75 \; \times \qquad n \qquad = 150 \qquad \text{Write a number sentence.}$$

$$\frac{0.75 \times n}{0.75} = \frac{150}{0.75} \qquad \text{Solve.}$$

$$n = 200$$

Write a number sentence and solve.

1. 40% of what number is 8?

   Number sentence _____ $n =$ _____

2. 7% of what number is 14?

   Number sentence _____ $n =$ _____

3. 25% of what number is 6?

   Number sentence _____ $n =$ _____

4. 15% of my allowance is $9. What is my allowance?

   Number sentence _____ $n =$ _____

5. 8% of the cost of new jeans is $4.80 for sales tax. What is the price of the jeans?

   Number sentence _____ $n =$ _____

6. The sale price of a jacket is $90. This price is 75% of the original price.
   What was the original price?

   Number sentence _____ $n =$ _____

Solve each problem.

7. 18% of $n = 9$; $n =$ _____   8. 20% of $n = 35$; $n =$ _____

9. 84% of $n = 126$; $n =$ _____   10. 12% of $n = 9$; $n =$ _____

11. 30% of $n = 45$; $n =$ _____   12. 90% of $n - 27$; $n =$ _____

13. 12% of $n = 72$; $n =$ _____   14. 46% of $n = 115$; $n =$ _____

15. 38% of $n = 209$; $n =$ _____   16. 7% of $n = 63$; $n =$ _____

17. 44% of $n = 11$; $n =$ _____   18. 25% of $n = 74$; $n =$ _____

Basic Skills Practice

Finding a Number When a Percent of It Is Known, Part 2

Read each question and circle the best answer.

1. 30% of what number is 3?

   A 9
   B 10
   C 90
   D 100

2. 32% of what number is 8?

   F 4
   G 8
   H 25
   I 256

3. Angel knows that 60% of a number is 90. What is the number?

   A 150
   B 300
   C 600
   D 666

4. I am thinking of a number. Two percent of the number is 16. What is the number?

   F 8
   G 80
   H 160
   I 800

Solve each problem.

5. Cleo the cat had kittens. Forty percent, or 2, of the kittens were female. How many kittens did Cleo have? _____

6. 80% of what number is 64? _____

7. When Coach Hamilton played 50% of his team, there were 11 players on the field. How many players does Coach Hamilton have on his team? _____

8. Arthur ate 12 pieces of candy. This was 25% of the package. How many pieces of candy were in the package? _____

9. The city police were using radar in the school zone this morning. Most of the tickets they wrote were for drivers going 40%, or 10 mph, over the speed limit. What is the speed limit in the school zone? _____

10. 53% of what number is 106? _____

11. 38% of what number is 19? _____

12. During the summer, a room at the beach is 28%, or $28, more expensive than during the fall. How much is a room at the beach during the fall? _____

Basic Skills Practice

Cumulative Review

Read each question and circle the best answer.

1. Of the 240 students in Rashel's 11th grade class, 60 live in apartments. What percent of the students live in apartments?

   A 10%

   B 25%

   C 60%

   D 180%

2. Angie sold 60 of the 75 cookies she brought to the class bake sale. What percent of the cookies did she sell?

   F 60%

   G 75%

   H 80%

   I 90%

3. Mario went to the movies, where his ticket cost $4.50. He also bought some candy for $2.00, popcorn for $2.50, and a jumbo soft drink for $3.75. How much money did Mario spend?

   A $13.75

   B $12.75

   C $12.25

   D $10.75

4. Of every 100 biography books in the school library, 8% of them are about sports figures. What decimal is equivalent to this percent?

   F 1.8

   G 0.88

   H 0.8

   I 0.08

5. 38% written as a decimal is—

   A 0.038

   B 0.38

   C 3.8

   D 38.0

6. 36% of what number is 72?

   F 36

   G 72

   H 200

   I 300

7. $\frac{27}{100}$ written as a percent is—

   A 0.27%

   B 2.7%

   C 27%

   D 270%

Solve each problem.

8. 54% of what number is 162?   _____

9. What is $\frac{3}{5}$ written as a percent?   _____

10. What is 38% of 95?   _____

11. What is 6.7×2.8?   _____

Basic Skills Practice

Using Proportions to Solve Percent Problems, Part 1

You can solve percent problems by using proportions.

Example 1: What number is 65% of 80?
This means: what number out of 80 equals 65 out of 100?

Rewrite as: $\dfrac{n}{80} = \dfrac{65}{100}$

Cross multiply: $100n = 5200$

Solve for n: $\dfrac{100n}{100} = \dfrac{5200}{100}$

$n = 52$

So 65% of 80 is 52.

Example 2: 9 is what percent of 12?
This means: what part of 100 equals 9 out of 12?

Rewrite as: $\dfrac{n}{100} = \dfrac{9}{12}$

$12n = 900$

$n = 75$

So 9 is 75% of 12.

Example 3: 5% of what number is 10?
This means: 5 out of 100 equals 10 out of what number?

$\dfrac{5}{100} = \dfrac{10}{n}$

$5n = 1000$

$n = 200$

So 5% of 200 is 10.

Solve by using a proportion.

1. What number is 50% of 48? _____

2. 20 is 50% of what number? _____

3. 25% of 12 is what number? _____

4. 8 is what percent of 80? _____

5. What percent of 15 is 9? _____

6. 8% of what number is 4? _____

7. 20% of what number is 5? _____

8. What percent of 20 is 6? _____

9. Gene ate 6 out of 8 slices of pizza. What percent of the pizza did
   Gene eat?   _____

10. Ten percent of the class left. Three students left. How many students
    are in the class?   _____

11. Out of 80 questions on the test, Laura answered 85% correctly. How
    many questions did she answer correctly?   _____

Basic Skills Practice

Using Proportions to Solve Percent Problems, Part 2

Read each question and circle the best answer.

1. Only 5% of the stores in the mall will not have a "Back to School" sale. If there are 80 stores in the mall, how many stores will have a "Back to School" sale?

   A 4
   B 40
   C 76
   D 78

2. What percent of 8 is 12?

   F 66%
   G 67%
   H 75%
   I 150%

3. Seventy-five percent of the 800 students at Thomas Jefferson High School voted for Band B to play at the end-of-the-year school dance. How many students voted for Band B?

   A 600
   B 625
   C 650
   D 700

4. How many students voted for a band other than Band B?

   F 100
   G 150
   H 175
   I 200

5. 10% of what number is 6?

   A 6
   B 10
   C 60
   D 600

6. Four out of 16 students completed the math test early. What percent of the students is this?

   F 16%
   G 25%
   H 40%
   I 50%

7. 24% of what number is 48?

   A 24
   B 48
   C 200
   D 248

8. What percent of 20 is 14?

   F 20%
   G 70%
   H 80%
   I 90%

9. What number is 75% of 40?

   A 30
   B 40
   C 75
   D 300

Solve.

10. Twenty-five percent of Melinda's family can water-ski on a slalom. If two people can slalom, how many people are there in Melinda's family? _____

Basic Skills Practice

Percent Increase and Decrease, Part 1

To find the percent increase or decrease:
*   Subtract to find the amount of increase or decrease. This number is the amount of change.
*   Write a ratio to show the amount of change as related to the original amount.
*   Rename the fraction as a percent.

Example 1:   The original amount is 50. The new amount is 75.

      • Subtract to find the amount of change. $75 - 50 = 25$

      • Write a ratio. $\dfrac{\text{change}}{\text{original}} = \dfrac{25}{50}$

      • Rename. $\dfrac{25}{50} = \dfrac{1}{2} = 50\%$ increase

Example 2:   The original amount is 60. The new amount is 54.

      • Subtract to find the amount of change. $60 - 54 = 6$

      • Write a ratio. $\dfrac{\text{change}}{\text{original}} = \dfrac{6}{60}$

      • Rename $\dfrac{6}{60} = \dfrac{1}{10} = 10\%$ decrease

Example 3:   Last week 80 students left campus for lunch. This week 100 students left campus for lunch. What is the percent increase?

      • Subtract to find the amount of change. $100 - 80 = 20$

      • Write a ratio. $\dfrac{\text{change}}{\text{original}} = \dfrac{20}{80}$

      • Rename. $\dfrac{20}{80} = \dfrac{1}{4} = 25\%$ increase

Find the percent change.

1.

Original amount	200	60	80
New amount	230	78	100
Percent of change			

2.

Original amount	40	60	300
New amount	34	36	270
Percent of change			

3. A radio that sold for $35 last year sells for $63 this year. What is the percent increase? _____

4. Compact discs are decreasing in cost. CDs that cost $20 last year can now be purchased for $15. What is the percent decrease? _____

5. Rafael lowered his cholesterol level from 250 to 150. What was the percent decrease in his cholesterol level? _____

6. Diana increased the number of clients in her photography studio from 12 per month to 18 per month. What was the percent increase? _____

7. During the after-Christmas sales, an artificial tree was on sale for $80. Before Christmas, the same tree was $200. What is the percent decrease? _____

Basic Skills Practice

Percent Increase and Decrease, Part 2

Read each question and circle the best answer.

1. Find the percent decrease if the original amount is 35 and the new amount is 14.

   A 60%
   B 50%
   C 35%
   D 14%

2. Adam usually scores 10 points per game. Last Thursday he had an exceptional game and scored 15 points. What percent increase was this?

   F 50%
   G 55%
   H 60%
   I 65%

3. To find percent change between an original amount, *A*, and a new amount, *B*, you should first—

   A write a ratio comparing *A* with *B*.
   B find the difference between *A* and *B*.
   C find the sum of *A* and *B*.
   D find the product of *A* and *B*.

4. A popular brand of jeans has had a price change from $45 a pair to $60 a pair. What is the approximate percent change?

   F 25% decrease
   G 25% increase
   H 33% decrease
   I 33% increase

5. Ginnette used to charge $10 for a haircut. She now charges $15. What is the percent increase?

   A 60%
   B 55%
   C 50%
   D 45%

If the original amount is 50 and the new amount is 25, what is—

6. the amount of change?

   F 20
   G 25
   H 50
   I 55

7. the percent change?

   A 20%
   B 25%
   C 50%
   D 55%

8. What is the percent increase if basketball shoes originally cost $100 and now cost $160?

   F 60%
   G 65%
   H 70%
   I 75%

9. Andrea found a game she wanted to buy for $55. When she returned to the store, the game was on sale for $33. What percent decrease was this?

   A 20%
   B 30%
   C 40%
   D 50%

10. The price of a sculpture is reduced from $600 to $450. What is the percent decrease?

   F 25%
   G 30%
   H 40%
   I 45%

Basic Skills Practice

Basic Skills Practice
Cumulative Review

Read each question and circle the best answer.

1. Melissa's goal is to run 25 miles each week. If she runs 5 miles on Sunday, 6 miles on Tuesday, 3 miles on Wednesday, and 6 miles on Thursday, what percent of her goal will she reach?

   A 20%

   B 25%

   C 50%

   D 80%

2. Attendance in an afterschool club increased from 16 to 20. What percent increase is this?

   F 16%

   G 20%

   H 25%

   I 50%

3. Sara has $80 to spend. If she spends 15% of her money, how much has Sara spent?

   A $12

   B $15

   C $25

   D $80

4. Eight is 25% of what number?

   F 8

   G 25

   H 48

   I 32

5. Jeff saved 12% on a coat that was marked $108. How much money did Jeff save?

   A $12.00

   B $12.96

   C $20.00

   D $25.00

6. What percent of 8 is 18?

   F 44%

   G 50%

   H 80%

   I 225%

7. Eric's paycheck for one pay period was $200. His next paycheck was $190. What percent decrease is this?

   A 5%

   B 10%

   C 15%

   D 20%

8. In the Austin Hot Rod Race, 14 of 56 cars did not finish the race. What percent represents the cars that did not finish the race?

   F 12.5%

   G 25%

   H 33%

   I 75%

Solve each problem.

9. Arthur has coupons to deliver. He has delivered 4%, or 10 coupons. How many coupons did Arthur have at the beginning? _____

10. Mary lost 5 lb last year. She wants to lose 15 lb this year. If Mary lost 3 lb in June, what percent of her desired weight loss for this year is this? _____

Basic Skills Practice

Using Ratios and Rates, Part 1

A ratio is a comparison of two numbers or quantities. Ratios may be written three different ways:

$$\frac{2}{7} \qquad 2 \text{ to } 7 \qquad 2 : 7$$

Ratios can be used to make predictions.

Example 1: Arthur makes 2 goals for every 7 attempts. Predict how many goals will he make in 35 attempts.

$$\text{Compare: } \frac{\text{goals}}{\text{attempts}} = \frac{2}{7}$$

Make equivalent ratios. (Making equivalent ratios is like making equivalent fractions.)

Goals	2	4	6	8	10
Attempts	7	14	21	28	35

We can expect Arthur to make 10 goals in 35 attempts.

Ratios can be used to determine how many times greater one number is than another.

Example 2: Richard is 16 years old. Martha is 4 years old. How many times older is Richard than Martha?

$$\text{Compare: } \frac{\text{Richard's age}}{\text{Martha's age}} = \frac{16}{4} \qquad \text{Divide: } \frac{16}{4} = 4$$

Richard is 4 times older than Martha.

Ratios can be used to compare amounts when the units of measure are the same.

Example 3: A punch recipe uses 4 cups of water and 16 ounces of orange concentrate. What is the ratio of water to concentrate?
Make sure the units are the same. Change 16 ounces to 2 cups.

$$\text{Compare: } \frac{\text{water}}{\text{concentrate}} = \frac{4 \text{ cups}}{2 \text{ cups}} = \frac{2 \text{ cups}}{1 \text{ cup}}$$

The ratio of water to concentrate is 2 cups of water to 1 cup of concentrate.

A ratio that compares two different kinds of quantities is called a rate.

$$\text{Examples of rates: } \frac{\text{miles}}{\text{hour}}, \frac{\text{miles}}{\text{gallon}}$$

When the rate has a denominator of 1, it is called a unit rate.

Example 4: A car travels 300 miles on 12 gallons of gasoline. What is the unit rate in miles per gallon (mpg)?
A unit rate compares a quantity to a unit of one.

$$\text{Compare: } \frac{\text{miles}}{\text{gallons}} = \frac{300}{12} = 25 \text{ mpg}$$

Write three equivalent ratios for each given ratio.

1. $\frac{5}{6}$ _____

2. $\frac{2}{3}$ _____

Write a ratio that compares each quantity.

3. the number of vowels to the number of consonants in the alphabet _____

4. the number of months that end in y to the total number of months in a year _____

Basic Skills Practice

Using Ratios and Rates, Part 2

Read each question and circle the best answer.

1. In Angie's class, 12 out of 16 students received a B on the test. Which ratio compares students who made a B with the total number of students?

  A $\frac{1}{8}$

  B $\frac{1}{4}$

  C $\frac{2}{5}$

  D $\frac{3}{4}$

2. Which ratios are equivalent to 6 : 8?

  F 7 : 9 and 8 : 10

  G 5 : 6 and 4 : 5

  H 3 : 4 and 12 : 16

  I 2 : 3 and 24 : 36

3. Coach Hamilton's team won 6 out of 12 games. Write in simplest form the ratio of games won to games played.

  A $\frac{1}{2}$

  B $\frac{3}{4}$

  C $\frac{5}{8}$

  D $\frac{5}{6}$

4. Lucille Ball's autograph is worth $75 to an autograph collector. President Truman's is worth $40, and Hillary Clinton's is worth $100. Which ratio compares the price of Lucille Ball's autograph with that of Hillary Clinton's?

  F 2:5

  G 3:4

  H 7:8

  I 9:10

5. The unit rate for 16 miles in 4 hours is—

  A 1 mile per hour

  B 2 miles per hour

  C 3 miles per hour

  D 4 miles per hour

6. If 36 balloons make 3 bunches, how many balloons make 1 bunch?

  F 24 balloons

  G 18 balloons

  H 12 balloons

  I 6 balloons

7. Misha traveled 233.2 miles in 4.4 hours. At what rate was Misha traveling?

  A 44 mph

  B 48 mph

  C 53 mph

  D 56 mph

8. A shirt factory uses spools of red thread for monograms on blue shirts. If one spool of thread will monogram 25 shirts, how many shirts would you expect 6 spools of thread to monogram?

  F 175 shirts

  G 150 shirts

  H 125 shirts

  I 100 shirts

9. It takes 2 ounces of syrup to sweeten a small snow cone. How many small snow cones can be sweetened with a 32-ounce bottle of syrup?

  A 8

  B 16

  C 32

  D 64

Basic Skills Practice

Using Proportions, Part 1

A proportion is an equation stating that two ratios are equal.

$$1:2 \text{ and } 4:8 \text{ form the proportion } \frac{1}{2} = \frac{4}{8}$$

Cross products can be used to tell if two ratios form a proportion.

$$\frac{1}{2} = \frac{4}{8} \quad 2 \times 4 = 8 \quad 1 \times 8 = 8 \quad \textit{Yes} \qquad\qquad \frac{1}{3} = \frac{4}{5} \quad 3 \times 4 = 12 \quad 1 \times 5 = 5 \quad \textit{No}$$

Cross products can be used to find the missing term in a proportion. $\quad \dfrac{18}{n} = \dfrac{6}{3}$

Example 1: Find the value of n in $\dfrac{18}{n} = \dfrac{6}{3}$.

$$\text{Use cross products: } 6 \times n = 18 \times 3$$
$$6n = 54$$
$$n = 9$$

Proportions can help solve problems.

Example 2: What is the cost of a dozen oranges if 3 oranges cost 99 cents?

$$\frac{3}{99} = \frac{12}{n}$$
$$3n = 1188$$
$$n = 396$$

12 oranges cost 396 cents or $3.96

Does each pair of ratios form a proportion? Write *yes* or *no*.

1. $\dfrac{3}{9}, \dfrac{6}{18}$ _____

2. $\dfrac{9}{10}, \dfrac{18}{30}$ _____

3. $\dfrac{1}{2}, \dfrac{50}{100}$ _____

4. $\dfrac{10}{20}, \dfrac{30}{40}$ _____

5. $\dfrac{55}{121}, \dfrac{5}{11}$ _____

6. $\dfrac{0.4}{2.3}, \dfrac{1.6}{9.2}$ _____

Solve each equation for *n*.

7. $\dfrac{48}{n} = \dfrac{4}{7}$ _____

8. $\dfrac{9}{24} = \dfrac{n}{48}$ _____

9. $\dfrac{4}{18} = \dfrac{6}{n}$ _____

10. $\dfrac{n}{55} = \dfrac{18}{22}$ _____

11. $\dfrac{5.1}{n} = \dfrac{1.7}{2}$ _____

12. $\dfrac{16}{34} = \dfrac{n}{1.7}$ _____

Solve by using a proportion.

13. Joe's favorite flavor of frozen yogurt is chocolate fudge. There are 65 calories in 2 ounces of chocolate fudge frozen yogurt. How many calories are there in 10 ounces? _____

14. One roll of gift wrap will wrap 4 shirt boxes. How many rolls will be needed to wrap 24 shirt boxes? _____

15. Elissa drives 164 miles in 4 hours. At that rate, how many miles will she travel in 6.5 hours? _____

Basic Skills Practice

Using Proportions, Part 2

Read each question and circle the best answer.

1. Which pair of ratios does not form a proportion?

   A $\frac{1}{4}, \frac{4}{16}$

   B $\frac{3}{5}, \frac{9}{15}$

   C $\frac{2}{3}, \frac{4}{9}$

   D $\frac{10}{15}, \frac{4}{6}$

2. Find the number to replace n in the proportion $\frac{3}{4} = \frac{18}{n}$.

   F 12
   G 16
   H 20
   I 24

3. Apples are being sold in a grocery store at 3 for 75 cents. How many apples can you buy for $2.25?

   A 8
   B 9
   C 10
   D 11

4. The ratio of the width of a banner to its length is 11 to 19. If the banner is 88 centimeters wide, what is its length?

   F 75 cm
   G 100 cm
   H 125 cm
   I 152 cm

5. Which equation could be used to find the missing term in the proportion $\frac{n}{4} = \frac{9}{2}$?

   A $4n = 18$
   B $9n = 8$
   C $2n = 36$
   D $2n = 13$

6. Compare the following proportions. Which one is true?

   F $\frac{2\text{ cans}}{\$1.52} = \frac{5\text{ cans}}{\$3.80}$

   G $\frac{168\text{ pens}}{7\text{ boxes}} = \frac{184\text{ pens}}{8\text{ boxes}}$

   H $\frac{300\text{ min}}{5\text{ tapes}} = \frac{350\text{ min}}{7\text{ tapes}}$

   I $\frac{360\text{ seats}}{8\text{ rows}} = \frac{441\text{ seats}}{9\text{ rows}}$

7. Thomas scores an average of 6 out of every 10 shots. How many shots should he score in 100 attempts?

   A 80
   B 70
   C 60
   D 50

8. At Lubbock High School the ratio of girls to boys is 2 to 3. If 630 boys attend the school, how many girls attend the school?

   F 240
   G 420
   H 460
   I 500

Basic Skills Practice

Cumulative Review

Read each question and circle the best answer.

1. Which ratio is not equivalent to the ratio 4:5?

   A 6:10

   B 12:15

   C 16:20

   D 32:40

2. Linda, Lana, and Kelly shared a pepperoni pizza. Linda ate $\frac{1}{3}$ of the pizza. Lana ate $\frac{3}{9}$ of the pizza, while Kelly claims to have eaten $\frac{9}{27}$ of the pizza. What conclusion can be drawn from this information?

   F Linda ate the most pizza.

   G Lana ate more than Linda but less than Kelly.

   H Kelly ate the least amount of pizza.

   I They all ate the same amount of pizza.

3. The ratio 48 to 24 expressed in simplest form is _____ to 1.

   A 5

   B 4

   C 2

   D 1

4. The leading team in the National League won 0.614 of its games. What percent of the games did this team win?

   F 0.614%

   G 6.14%

   H 61.4%

   I 614%

5. The school store decided to sell T-shirts featuring the school mascot. They ordered 4 dozen shirts and sold 20 of them the first morning. What is the ratio of the number of shirts sold the first morning to the number of shirts ordered?

   A $\frac{4}{20}$

   B $\frac{20}{4}$

   C $\frac{20}{48}$

   D $\frac{48}{20}$

6. Solve $18 - 9 \div 3 \times 2$.

   F 3

   G 6

   H 12

   I 15

7. Among the fractions $\frac{1}{2}, \frac{7}{25}, \frac{17}{30}$, and $\frac{39}{50}$, which fraction is greater than $\frac{3}{5}$?

   A $\frac{1}{2}$

   B $\frac{7}{25}$

   C $\frac{17}{30}$

   D $\frac{39}{50}$

8. If 2 doughnuts cost 75 cents, how much would 6 doughnuts cost?

   F $1.50

   G $3.00

   H $2.25

   I $6.00

Basic Skills Practice

Points, Lines, and Planes, Part 1

All geometric figures are made up of a set of points. Drawings or symbols represent these figures.

Definition	Drawing	Symbol	Definition	Drawing	Symbol
A **point** is a location in space with no size.	• Z	Z	A **line segment** is a part of a line with two endpoints.	•———• Q　　R	$\overline{QR}$ or $\overline{RQ}$
A **plane** is a flat surface extending endlessly in all directions.	plane with A, B, C / plane ℜ	plane ABC plane $\mathfrak{R}$	A **ray** is a piece of a line with one endpoint.	•—•——→ M　N	$\overrightarrow{MN}$
A **line** is a set of points in a straight path that extends endlessly in opposite directions.	←•——•→ A　B ←———→ t	$\overleftrightarrow{AB}$ or $\overleftrightarrow{BA}$ t	An **angle** is made up of two rays with the same endpoint, called a **vertex**.	X ↗ Y •—•→ Z	$\angle XYZ$ or $\angle ZYX$ or $\angle Y$ (the vertex)

Identify each symbol.

1. $\overrightarrow{JS}$ _____

2. $\angle Y$ _____

3. $\overleftrightarrow{DF}$ _____

4. $\overrightarrow{AN}$ _____

Identify and name each figure.

5.  _____

6. 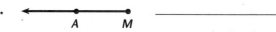 _____

7. 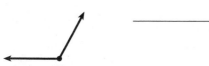 _____

8.  _____

9. 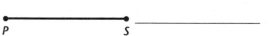 _____

10. • R _____

11. _____

12. _____

Basic Skills Practice

Points, Lines, and Planes, Part 2

Read each question and circle the best answer.

1. What shape has two sets of parallel lines that do *not* intersect at 90° angles?

   A rectangle
   B trapezoid
   C parallelogram
   D square

2. At what point do lines *b* and *d* meet?

   F *F*
   G *D*
   H *E*
   I *G*

   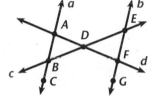

3. The darkened lines represent—

   A parallel lines
   B perpendicular lines
   C intersecting lines
   D skew lines

   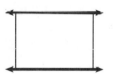

4. The vertex of a pyramid is—

   F a line
   G a point
   H a line segment
   I a ray

5. The darkened portion of this figure represents—

   A a line
   B a ray
   C an angle
   D a line segment

   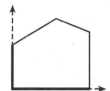

6. The edge of a rectangular prism is—

   A an angle
   B a rectangle
   C a line segment
   D a triangle

7. Angle *KLJ* includes—

   F line *KI*
   G line *KL*
   H line segment *KI*
   I line segment *KL*

   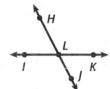

8. In this figure, what is the total number of rays with endpoint *D*?

   F 2
   G 3
   H 4
   I 6

   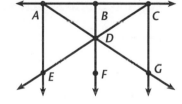

Use the figures to answer each question.

9. How many line segments are in this drawing?

   _____

   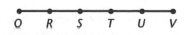

10. Name the angle in four ways.

    _____

Basic Skills Practice

Angles and Angle Relationships, Part 1

You can classify angles by their measures. Study the table below.

Angle or angle pairs	Name	Definition
or	acute angle	An **acute angle** is an angle that measures more than 0° and less than 90°.
or	right angle	A **right angle** measures exactly 90°.
or	obtuse angle	An **obtuse angle** measures more than 90° and less than 180°.
	straight angle	A **straight angle** measures exactly 180°.
30° T S 60° Q R	complementary angles	Two angles are **complementary** if the sum of their measures is 90°.
F 45° 135° E H G	supplementary angles	Two angles are **supplementary** if the sum of their measures is 180°.
B A C D	vertical angles	**Vertical angles** are pairs of opposite angles formed by intersecting lines. They have equal measures. Angles *A* and *C* are vertical angles.
B A C D	adjacent angles	**Adjacent angles** have a common vertex and side, and no common interior points. Angles *A* and *B* are adjacent angles.

The sum of the measures of the angles inside a triangle equals 180°.

Classify each angle as acute, right, obtuse, or straight.

1. _____

2. _____

3. 35° _____

4. 180° _____

5. 90° _____

Tell whether each pair of angles is complementary, supplementary, or neither.

6.  _____

7. _____

8. 115°, 65° _____

9. 45°, 65° _____

10. 55°, 25° _____

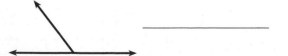

Basic Skills Practice

Angles and Angle Relationships, Part 2

Read each question and circle the best answer.

1. An isosceles triangle has two congruent sides and two congruent angles. The vertex angle of this isosceles triangle is 50° and its sides are 7 units. What is the measure of each base angle?

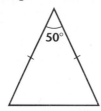

   A 20°

   B 50°

   C 65°

   D 80°

2. Triangle *BCD* is an isosceles triangle. What is the measure of ∠*CDB* if ∠*ABC* is 130°?

   F 50°

   G 60°

   H 90°

   I 130°

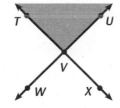

3. The shaded section of this figure can best be identified as—

   A ∠*V*

   B ∠*UTV*

   C ∠*TVU*

   D ∠*TUV*

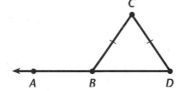

4. Which angle has point *L* as its vertex?

   F ∠*KLM*

   G ∠*LNP*

   H ∠*INL*

   I ∠*LIJ*

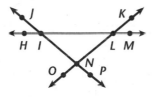

5. If ∠1 measures 70°, what is the measure of ∠2?

   A 20°

   B 70°

   C 90°

   D 110°

6. What is the measure of each unmarked angle in this isosceles right triangle?

   F 30°

   G 45°

   H 60°

   I 90°

7. What is the missing angle measurement in this triangle?

   A 28°

   B 38°

   C 53°

   D 65°

8. The measure of ∠1 is 110°. What is the measure of ∠*K*?

   F 60°

   G 70°

   H 80°

   I 90°

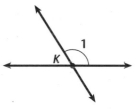

Solve each problem.

9. Angle *M* and angle *N* are supplementary angles. If angle *M* measures 35°, what is the measure of angle *N*? _____

10. Angle *J* and angle *K* are complementary angles. If angle *J* measures 27°, what is the measure of angle *K*? _____

Basic Skills Practice

Parallel Lines in Geometry, Part 1

Line *h* is **parallel** to line *f*. Line *l* is a **transversal**.
A transversal is a line that intersects two parallel lines.

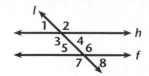

Alternate interior angles are the pairs of angles that are inside the parallel lines and on opposite sides of the transversal. Alternate interior angles are congruent. Examples: ∠3 ≅ ∠6 and ∠4 ≅ ∠5

Alternate exterior angles are the pairs of angles that are outside the parallel lines and on opposite sides of the transversal. Alternate exterior angles are congruent. Examples: ∠1 ≅ ∠8 and ∠2 ≅ ∠7

Corresponding angles are the pairs of angles that are in the same position at the intersection of each parallel line and the transversal. Corresponding angles are also congruent. Examples: ∠4 ≅ ∠8 and ∠1 ≅ ∠5

Use the figure to complete Exercises 1–6.

Find the alternate interior angles.

1. ∠3 and ∠ _____   2. ∠4 and ∠ _____

Find the alternate exterior angles.

3. ∠1 and ∠ _____   4. ∠2 and ∠ _____

Find the corresponding angles.

5. ∠3 and ∠ _____   6. ∠1 and ∠ _____

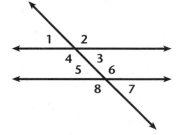

Use the diagram to find the measure of each angle.

7. ∠10 _____   8. ∠12 _____   9. ∠13 _____

10. ∠14 _____   11. ∠15 _____   12. ∠17 _____

In the figure, lines *d* and *f* are parallel, and lines *q* and *s* are parallel. Find the measure of each angle if the measure of ∠*b* = 47°.

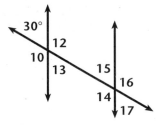

13. ∠*a* _____   14. ∠*d* _____   15. ∠*j* _____

16. ∠*h* _____   17. ∠*l* _____   18. ∠*i* _____

19. ∠*f* _____   20. ∠*m* _____

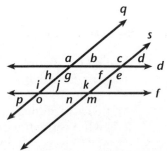

Basic Skills Practice

Parallel Lines in Geometry, Part 2

Read each question and circle the best answer.

1. Lines *a* and *b* are parallel. If the measure of ∠1 is 140°, what is the measure of ∠3?

    A 140°

    B 90°

    C 40°

    D 10°

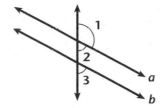

2. Lines *f* and *g* are parallel. What is the measure of ∠*t*?

    F 128°

    G 62°

    H 52°

    I 22°

3. Lines *b* and *c* are parallel. Which angle does not share the same measurement as ∠3?

    A ∠1

    B ∠5

    C ∠7

    D ∠8

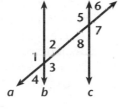

4. Lines *q* and *r* are parallel. If ∠A measures 38°, what is the measure of ∠B?

    F 12°

    G 52°

    H 62°

    I 142°

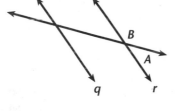

Use the figure to answer Exercises 5–7.

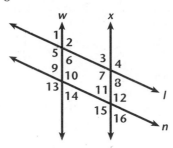

5. Lines *w* and *x* are parallel. Lines *l* and *n* are also parallel. Which of the following has the same measure as ∠10?

    A ∠4

    B ∠6

    C ∠9

    D ∠16

6. If ∠3 measures 70°, ∠11 measures—

    F 20°

    G 50°

    H 70°

    I 110°

7. If ∠10 measures 110°, ∠9 measures—

    A 40°

    B 50°

    C 60°

    D 70°

Lines *a* and *b* are parallel. Use the figure to answer Exercises 8–9.

8. If the measure of ∠1 is 50°, what is the measure of ∠2?

_____

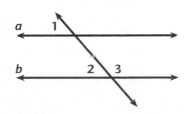

9. If ∠3 measures 130°, what is the measure of ∠2?

_____

Basic Skills Practice

Cumulative Review

Read each question and circle the best answer.

1. The vertex angle in this isosceles triangle measures 20°. What is the measure of each base angle?

   A 20°

   B 35°

   C 60°

   D 80°

2. In this figure, what is the total number of line segments in which the endpoints are at *K, L, M,* or *N*?

   F 8

   G 6

   H 5

   I 3

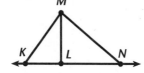

3. Lines *r* and *s* are parallel. If ∠1 measures 115°, what is the measure of ∠2?

   A 115°

   B 105°

   C 85°

   D 65°

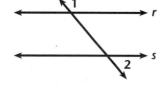

4. What is the missing angle measurement in this triangle?

   F 22°

   G 32°

   H 42°

   I 52°

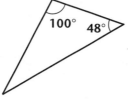

5. If ∠1 measures 125°, what is the measure of ∠2?

   A 55°

   B 65°

   C 125°

   D 135°

6. Points *T, U, V,* and *W* are on the same line segment. Points *X, V, Y,* and *Z* are on the same line segment. Which statement is true?

   F $TV + UV = TU$

   G $TV - UV = TU$

   H $XV + YZ = XZ$

   I $ZX - YV = ZY$

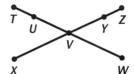

7. A daily allowance of 60 milligrams of vitamin C is recommended for 18-year-olds. If 15 grapes have about 2 milligrams of vitamin C, which proportion could be used to find *G*, the number of grapes needed to provide 60 milligrams?

   A $\dfrac{G}{15} = \dfrac{2}{60}$

   B $\dfrac{18}{60} = \dfrac{15}{G}$

   C $\dfrac{G}{2} = \dfrac{15}{60}$

   D $\dfrac{15}{2} = \dfrac{G}{60}$

Solve each problem.

8. Frank bought a pair of basketball shoes for $44.99. The original price was $59.95. How much money did he save? _____

9. What is the missing number in this pattern: 1, 4, _____, 16, 25, 36? _____

Basic Skills Practice

Statistics: Mean and Range, Part 1

The mean or average is the sum of the data divided by the number of pieces of data.

The range of a set of data is the difference between the greatest value and the least value.

Example: Jonathan has taken 4 quizzes in math in this six-weeks period. His scores are 90, 85, 73, and 80. Find the mean and the range of Jonathan's scores.

Find the mean.	Find the range.
1. Add the scores.	Subtract the least score (73)
$\quad$ 90 + 85 + 80 + 73 = 328	from the greatest score (90).
2. Divide by the number of scores.	90 − 73 = 17
$\quad$ 328 ÷ 4 = 82	
The mean is 82.	The range is 17.

Find the mean of each group of numbers.

1. 98, 72, 80, 84, 91 _____

2. 3, 0.8, 2.6, 8.4, 9.2 _____

3. 45, 31, 42, 67, 60 _____

4. 5, 1.5, 3.4, 4, 6.1 _____

Find the range of each group of numbers.

5. 56, 68, 57, 72, 70 _____

6. 6.4, 5.7, 4.3, 6.3 _____

7. 40, 45, 47, 52, 65 _____

8. 2.7, 3.2, 4.3, 5.8 _____

Solve each problem.

9. Leo took 6 math tests. His scores were 69, 89, 92, 94, 88, and 90. Find the range and the mean for Leo's scores. _____

10. Juanita works as a volunteer. In the last 4 weeks she worked 10 hours, 14 hours, 12 hours, and 4 hours. On the average, how many hours per week did she work? _____

11. Daniel needs an average of at least 20 points to enter the final round of a contest. So far he has accumulated 104 points in 5 events. How many points does he need in the last event to qualify for the final round? _____

12. Emilie and her brother Joel went on a bicycling trip to Colorado. They traveled a total of 150 miles in 5 days. What was the average number of miles traveled per day? _____

Basic Skills Practice

Statistics: Mean and Range, Part 2

Read each question and circle the best answer.

1. In the last 4 basketball games, Angela scored 12, 14, 10, and 16 points. What is her mean for the last 4 games?

   A 10 points

   B 11 points

   C 12 points

   D 13 points

2. Melissa and Angie spent $24, $30, $20, and $18 on groceries for 4 weeks. What is the average amount spent each week?

   F $23

   G $22

   H $20

   I $18

3. The ticket prices available for a Texas Rangers' baseball game are $14, $12, $24, $15, $8, and $11. Find the range.

   A $18

   B $16

   C $12

   D $10

4. The average daily temperatures for a week are 86, 84, 78, 92, 87, 91, and 77. Which measure would best describe the temperature variations for this week?

   F the total

   G the range

   H the mean

   I the highest temperature

5. Four friends went to shop for school supplies. Their expenses were $23.67, $19.24, $31.86, and $25.75. What is the average amount spent on supplies?

   A $23.67

   B $24.82

   C $25.13

   D $27.55

6. The Johnson family is planning a road trip from Amarillo to San Antonio. The distance from Amarillo to San Antonio is 515 miles. The Johnsons traveled 53 miles the first hour, 57 miles the second hour, and 61 miles the third hour. What is the average number of miles traveled per hour?

   F 56 mi

   G 57 mi

   H 58 mi

   I 59 mi

7. Six different surveyors made these measurements to determine the height of a mountain.

8990 feet	8992 feet	8999 feet
9002 feet	9005 feet	9026 feet

   Find the range of the measurements.

   A 25 ft

   B 26 ft

   C 36 ft

   D 42 ft

Solve each problem.

8. The 5 tenth-grade classes have 22, 21, 20, 20, and 22 students. What is the average number of students per class? _____

9. Find the mean of 26, 34, and 27. _____

Basic Skills Practice

Statistics: Mean and Mode, Part 1

Recall that the mean or average is the sum of the data divided by the number of pieces of data. The mode is the piece of data that appears most often in a data list. There may be one mode, more than one mode, or no mode. The mode is useful when data pieces are not numerical.

Example: Find the mode of the following data lists:

1. 95, 80, 92, 91, 98, **94, 94** The mode is 94 because 94 appears most often.

2. **4, 5, 5, 4,** 6 The modes are 4 and 5 because 4 and 5 appear most often.

3. 99, 100, 101 There is no mode because no number appears more often than another.

The chart shows the hours that Sara spent working in a flower shop for a one-week period.

Sunday	Monday	Tuesday	Wednesday	Thursday	Friday	Saturday
2	7	6.5	8.5	7	8	3

1. What is the mode of the number of hours that Sara worked? _____

2. What is the mean (average) number of hours that Sara worked? _____

Find the mean and mode for the following data lists. If necessary, round to the nearest tenth.

3. 2, 5, 6, 7, 4, 6   Mean _____   Mode _____

4. 8, 5, 12, 3, 4, 5, 14, 5   Mean _____   Mode _____

5. 80, 65, 78, 72   Mean _____   Mode _____

6. 2, 3, 2, 2, 5, 3, 3   Mean _____   Mode _____

7. 1, 2, 3, 4, 5, 6   Mean _____   Mode _____

In the 100-yard dash, 5 runners had these times: 10.4 seconds, 11.2 seconds, 10.6 seconds, 12.1 seconds, and 11.2 seconds.

8. What was the mean time? _____

9. What was the mode of the times? _____

10. What was the range of the times? _____

Basic Skills Practice

Statistics: Mean and Mode, Part 2

Read each question and circle the best answer.

1. The Lewisville Panthers' football scores for the first 4 games of the season are 17, 13, 19, and 12. To find the mean (average) of these scores, what should you do first?

   A Multiply the two greatest scores.
   B Subtract the least score from the greatest score.
   C Find the sum of all the scores.
   D Compare the two greatest scores.

2. Tuition costs for Julie's first 4 semesters at Texas Tech University were $462, $586, $323, and $473. What was the average cost of tuition per semester?

   F $461
   G $475
   H $500
   I $506

3. Which statement about mode is false?

   A There will always be a mode.
   B There may be one mode.
   C There may be more than one mode.
   D The mode is the data item that appears most often.

4. Which set of grades has the highest average?

   F 74, 80, 92, 82, 92
   G 74, 80, 74, 82, 85
   H 74, 80, 92, 85, 74
   I 74, 80, 70, 71, 80

5. Liliana's long-distance telephone bills were $145, $160, $115, and $120 for January through April. What was the average monthly amount that Liliana spent on long-distance calls?

   A $150
   B $140
   C $135
   D $130

6. What are the mean and mode of the numbers 2, 5, 6, 7, 4, and 6?

   F 6 and 4
   G 5 and 6
   H 4.5 and 5
   I 4 and 6

7. Which 7 numbers have a mean of 5?

   A 6, 3, 3, 6, 5, 6, 5
   B 5, 5, 2, 8, 6, 5, 3
   C 4, 5, 5, 4, 6, 6, 5
   D 3, 4, 5, 4, 6, 6, 3

8. In English Jena has test scores of 98, 72, 80, 84, and 91. Which measure will give the most information about Jena's test scores?

   F the total of the scores
   G the range of the scores
   H the mode of the scores
   I the mean of the scores

Solve.

9. The ages of a volleyball team's players are 20, 21, 21, 24, and 25. What is the mode? _____

Basic Skills Practice

Cumulative Review

Read each question and circle the best answer.

Bowling Scores

Bowler	Game 1	Game 2	Game 3	Game 4
Gene	126	132	188	122
Matt	132	134	160	126
Brad	112	125	125	150
Nathan	120	144	138	138

1. What is Gene's average for the four games?

   A 123

   B 130

   C 142

   D 188

2. What is the range of Matt's scores?

   F 34

   G 30

   H 28

   I 20

3. Which bowler has the highest average?

   A Gene

   B Matt

   C Brad

   D Nathan

4. To determine the best bowler, what should you consider?

   F the mean

   G the range

   H the mode

   I the number of games bowled

5. Edith bought 10 large T-shirts and 8 medium T-shirts for her team. Write the ratio of large T-shirts to medium T-shirts in simplest form.

   A 8 to 10

   B 10 to 8

   C 4 to 5

   D 5 to 4

6. Which measure would be most appropriate to use when considering the favorite subject of students in your grade?

   F the mean

   G the range

   H the mode

   I the ratio

7. During an average week, Becca works 21 hours for every 28 hours that Juan works. Write the ratio of Becca's hours per week to Juan's hours per week in simplest form.

   A 21 to 28

   B 3 to 4

   C 4 to 3

   D 4 to 7

8. While on vacation for one week, the Hamiltons spent $352 on food, $220 on lodging, $80 on gas, and $20 on gifts. What is a reasonable estimate for the average cost of the vacation per day?

   F between $70 and $80

   G between $80 and $90

   H between $90 and $100

   I between $100 and $110

Basic Skills Practice

Squares and Square Roots, Part 1

When two equal numbers are multiplied, the product is called the square of the number.

Example 1: $3 \times 3 = 3^2 = 9$ 9 is the square of 3.
               $8 \times 8 = 8^2 = 64$ 64 is the square of 8.

Finding the square root of a product means finding one of its two equal factors. Positive numbers can have either two positive or two negative equal factors. For example, $4 = 2 \times 2$ or $4 = -2 \times -2$. The symbol $\sqrt{}$ is used to indicate the positive square root of a number.

Example 2: $\sqrt{16} = 4$ because $4 \times 4 = 16$
             $\sqrt{100} = 10$ because $10 \times 10 = 100$
             $\sqrt{\frac{4}{9}} = \frac{2}{3}$ because $\frac{2}{3} \times \frac{2}{3} = \frac{4}{9}$

Squares and square roots are used in problem solving.

Example 3: Find the length of the side of a square whose area is 36 cm^2.
                  $\sqrt{36} = 6$
    The length of the side is 6 cm.

Example 4: Find the area of a square if the length of the side is 5 m.
             $5^2 = 5 \times 5 = 25$ m^2.
             The area of the square is 25 m^2.

Find the square of each number.

1. 4 _____

2. 7 _____

3. 23 _____

Solve.

4. 11^2 _____

5. 20^2 _____

6. 51^2 _____

Find the square root of each number.

7. 25 _____

8. 1 _____

9. 81 _____

Solve.

10. $\sqrt{36}$ _____

11. $\sqrt{100}$ _____

12. $\sqrt{\frac{9}{16}}$ _____

Basic Skills Practice

Squares and Square Roots, Part 2

Read each question and circle the best answer.

1. To find the square of the number 19, you should —

   A multiply 19×2

   B add $19 + 19$

   C multiply 19×19

   D divide 19 by 2

2. A perfect square is a number whose square root is a whole number. Which of these is not a perfect square?

   F 16

   G 24

   H 36

   I 49

3. Melissa has to buy carpeting for her son's playroom. The room is square, and the side measures 20 feet. What should she do to calculate the amount of carpet that she needs?

   A Multiply 20×4.

   B Multiply 20×20.

   C Divide 20 by 4.

   D Find the square root of 20.

4. Find the length of the side of a square whose area is 121 cm^2.

   F 11 cm

   G 10 cm

   H 9 cm

   I 8 cm

5. The formula $M = 1.22 \times \sqrt{A}$ is used to predict how far a passenger in a plane can see in miles (M) when the altitude (A) is in feet. If an airplane is flying at an altitude of 10,000 feet, how far can a passenger see?

   A $M = 1.22 \times \sqrt{10}$

   B $M = 1.22 \times \sqrt{100}$

   C $M = 1.22 \times \sqrt{1000}$

   D $M = 1.22 \times \sqrt{10,000}$

6. The square target for a parachute jump has an area of 225 square feet. How long is each side of the target?

   F 25 ft

   G 22 ft

   H 20 ft

   I 15 ft

7. Find the value of $\sqrt{144}$.

   A 12

   B 16

   C 36

   D 72

8. What is the value of 9^2?

   F 18

   G 81

   H 215

   I 512

Solve each problem.

9. A square has sides that measure 16 cm. Find the area of the square. _____

10. What is the square of 17? _____

Basic Skills Practice

The "Pythagorean" Right-Triangle Theorem, Part 1

If **a** and **b** are the lengths of the legs or sides of a right triangle and **c** is the length of the hypotenuse (the longest side), then $a^2 + b^2 = c^2$.

$a = 3$, $b = 4$, and $c = 5$

$3^2 + 4^2 = 5^2$
$9 + 16 = 25$
$25 = 25$

This is called the "Pythagorean" Right-Triangle Theorem. If the lengths of any two sides of a right triangle are known, the third side can be found by using this theorem.

Example 1: If $a = 6$ and $b = 8$, find the value of c.
1. Replace and compute: $a^2 + b^2 = c^2$
$6^2 + 8^2 = c^2$
$36 + 64 = c^2$
$100 = c^2$
2. Solve by taking the square root.
$\sqrt{100} = c$
$10 = c$

The length of side c is 10.

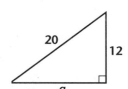

Example 2: If $b = 12$ and $c = 20$, find the value of a.
1. Replace and compute:

$a^2 + b^2 = c^2$
$a^2 + 12^2 = 20^2$
$a^2 + 144 = 400$
$a^2 + 144 - 144 = 400 - 144$
$a^2 = 256$
$a = 16$

The lengths of the sides of a triangle are given. Do the three sides form a right triangle? (Does $a^2 + b^2 = c^2$?) Answer *yes* or *no*.

1. $a = 9$ cm, $b = 11$ cm, and $c = 15$ cm _____

2. $a = 8$ cm, $b = 15$ cm, and $c = 17$ cm _____

Find the missing side of each right triangle. Use $a^2 + b^2 = c^2$.

3. $a = 12$ cm, $b = 16$ cm, and $c =$ _____

4. $a =$ _____, $b = 15$ m, and $c = 17$ m

5. Find x. _____

6. Find m. _____

7. Find h. _____

8. Find d. _____

Basic Skills Practice

The "Pythagorean" Right-Triangle Theorem, Part 2

Read each question and circle the best answer.

1. A right triangle has legs that measure 10 cm and 24 cm. Using the "Pythagorean" Right-Triangle Theorem, which equation would give the length of the hypotenuse?

   A $100 + 576 = c^2$
   B $a^2 + 100 = 576$
   C $10 + 24 = c^2$
   D $576 + b^2 = 100$

2. A piece of wood that is shaped like a right triangle is used to prop open a heavy door to the school gym. If the lengths of the shortest sides are 5 inches and 12 inches, what is the length of the longest side?

   F 16 in.
   G 15 in.
   H 14 in.
   I 13 in.

3. The ladder to a playground slide and the slide itself form a right triangle with the ground. If the ladder is 9 feet high and the distance from the foot of the ladder to the end of the slide is 12 feet, what is the length of the slide?

   A 15 ft
   B 16 ft
   C 17 ft
   D 18 ft

4. Given a right triangle with sides a and b and a hypotenuse of 25 inches, which of the following equations is correct?

   F $a^2 + b^2 = 25$
   G $a^2 + 25 = c^2$
   H $25 + b^2 = a^2$
   I $a^2 + b^2 = 25^2$

Consider the lengths of the sides of each triangle. Write *yes* if the sides form a right triangle, and write *no* if they don't.

5. $a = 7$, $b = 24$, and $c = 25$ _____

6. $a = 3$, $b = 4$, and $c = 6$ _____

7. $a = 9$, $b = 16$, and $c = 31$ _____

8. $a = 5$, $b = 8$, and $c = 14$ _____

Find the missing length.

9. $a =$ _____

10. $c =$ _____

11. $b =$ _____

12. $d =$ _____

Basic Skills Practice

Cumulative Review

Read each question and circle the best answer.

1. State an equation that you can use to find the length of the hypotenuse of a right triangle if the sides measure 13 m and 9 m.

   A $169 + 81 = c^2$

   B $a^2 + 81 = 169$

   C $169 + b^2 = 81$

   D $12 + 9 = c^2$

2. Mrs. Shelton's class had a test on fractions. Twenty-five percent of the class need to take the test again. If there are 24 students in Mrs. Shelton's class, how many students need to take the test again?

   F 4

   G 6

   H 8

   I 12

3. A weather balloon is 8 km east and 5 km north of the weather station. About how far from the weather station is the weather balloon?

   A about 11 km

   B about 9 km

   C about 8 km

   D about 7 km

4. The first three perfect squares are 1, 4, 9. What are the next three perfect squares?

   F 4, 9, 16

   G 9, 16, 25

   H 16, 25, 36

   I 25, 36, 49

5. A chocolate chip cookie recipe calls for $1\frac{1}{3}$ cups of brown sugar and $\frac{3}{4}$ cup of granulated sugar. How many cups of sugar does the recipe contain?

   A $1\frac{1}{2}$ cups

   B $1\frac{4}{7}$ cups

   C 2 cups

   D $2\frac{1}{12}$ cups

6. Order the following numbers from least to greatest: 1695, 15,906, 1596, 1569.

   F 1596, 1569, 1695, 15,906

   G 15,906, 1596, 1569, 1695

   H 1569, 1596, 1695, 15,906

   I 15,906, 1695, 1596, 1569

Solve each problem.

7. What is 9.528 rounded to the nearest tenth? _____

8. A square piece of fabric has an area of 900 square inches. What is the length of each side? _____

9. Gina is helping her father paint their house. If Gina places the 10-foot ladder 6 feet from the house and leans it against the house, what is the distance from the bottom of the house to the top of the ladder? _____

Basic Skills Practice
Cumulative Review

Read each question and circle the best answer.

1. Find the missing numbers in the pattern.
   34, 30, 28, 24, _____, _____, 16, 12, _____

   A 20, 22, 8
   B 22, 18, 10
   C 22, 20, 18
   D 22, 20, 6

2. Which set of fractions represents equivalent amounts of pizza?

   F $\frac{2}{3}, \frac{3}{4}, \frac{4}{5}$

   G $\frac{3}{8}, \frac{5}{8}, \frac{6}{8}$

   H $\frac{3}{4}, \frac{6}{8}, \frac{12}{16}$

   I $\frac{5}{6}, \frac{5}{7}, \frac{5}{9}$

3. Andrea lives 4.8 km from the park where her soccer team practices each week. How many meters is this?

   A 4.8
   B 48
   C 480
   D 4800

4. Nicole went to the store to buy ice cream for the school party. She needed 8 one-gallon containers, but the store only had pints. How many pints of ice cream did she buy?

   F 16
   G 36
   H 64
   I 82

5. Which unit of measure would be the most appropriate to express the thickness of an envelope?

   A millimeter
   B centimeter
   C decimeter
   D meter

6. Sara pours three glasses of juice from a 6-cup juice pitcher. What does Sara need to know in order to determine how much juice is left in the pitcher?

   F how much juice is in the pitcher
   G what kind of juice is in the pitcher
   H how much each glass will hold
   I how much each glass weighs

Solve each problem.

7. If high school juniors spend $7\frac{1}{2}$ hours in school each day, excluding extracurricular activities, how many hours do high school juniors spend in school each week? _____

8. Multiply $\frac{3}{5} \times \frac{4}{9} \times \frac{10}{16}$. _____

9. Simplify $9 \div 3 + 4 \times 7 - 20 \div 5$. _____

Basic Skills Practice
Cumulative Review

Read each question and circle the best answer.

1. Which expression is equivalent to $6(2 + n)$?

   A $2 + 6n$

   B $12n$

   C $12 + 6n$

   D $12 + n$

2. The students in Mrs. King's class are collecting money for prom flowers. They have collected the following amounts: $23.75, $18.90, $30.50, and $27.25. How much money have the students collected?

   F $99.40

   G $100.40

   H $100.46

   I $110.25

3. Suppose that math textbooks cost $20.87 per book. If Mrs. King needs to order 8 books, what would the total cost be?

   A $120.06

   B $120.96

   C $156.96

   D $166.96

4. A customer buys three dozen bagels for $15.00. Sales tax is 8%. What is the total cost after tax?

   F $15.80

   G $16.20

   H $16.80

   I $17.20

5. The value of Patty's bike was $205 when she bought it. This year, its value is $164. What is the percent decrease in value?

   A 5%

   B 10%

   C 15%

   D 25%

   E Not Here

6. Jennifer's new car uses 6 gallons of gas to travel 240 miles. How many miles does the car travel on 1 gallon?

   F 30 mi

   G 40 mi

   H 45 mi

   I 50 mi

Solve each problem.

7. Sales tax in the Austin area is 8.25%. If Ryan buys 2 shirts for $25.96 each, how much is the sales tax? _____

8. A square has a perimeter of 18.4 centimeters. What is the length of its sides? _____

9. The water container that the football team keeps on the sidelines holds 6 gallons of water. If it now has $4\frac{1}{2}$ gallons of water, how many ***quarts*** of water need to be added to fill the container to capacity? _____

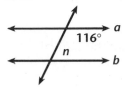

Basic Skills Practice

Cumulative Review

Read each question and circle the best answer.

1. Lines *a* and *b* are parallel. What is the value of *n*?

   116° *a*
   n *b*

   A 34°
   B 64°
   C 74°
   D 116°

2. Which proportion could be used to find the number that is 98% of 140?

   F $\dfrac{140}{n} = \dfrac{98}{100}$

   G $\dfrac{98}{n} = \dfrac{140}{100}$

   H $\dfrac{n}{140} = \dfrac{98}{100}$

   I $\dfrac{98}{140} = \dfrac{n}{100}$

3. What is the sum of $0.596 + 2.536$ rounded to the nearest tenth?

   A 3.132
   B 3.13
   C 3.1
   D 3.0

4. Jan uses 16 eggs to make her famous french toast for 60 people. How many eggs would she need to make french toast for 90 people?

   F 20
   G 22
   H 24
   I 26

5. Sally bought five items that were priced $3.98, $7.09, $13.14, $1.89, and $0.79. What is the most reasonable estimate for the total amount she paid?

   A Less than $20
   B Between $20 and $25
   C Between $25 and $30
   D More than $30

6. Kelly had a roll of red ribbon containing $10\frac{1}{2}$ yards of ribbon. She used $5\frac{3}{4}$ yards for party decorations. How much ribbon is left on the roll?

   F $4\frac{1}{2}$ yd

   G $4\frac{3}{4}$ yd

   H 5 yd

   I $16\frac{1}{4}$ yd

Solve each problem.

7. The approximate braking distance, in feet, necessary to stop a car is given by the formula $D = 0.05R^2$, where *R* is the speed of the car. What is the approximate braking distance for a car traveling 50 miles per hour?

   _____

8. Alberto is on a special diet. He can eat only 2100 calories a day, and only 20% of these can be calories from fat. How many calories from fat is Alberto allowed to eat?

   _____

Basic Skills Practice

Writing Equations and Inequalities, Part 1

Translating words into mathematical symbols can help solve problems. Let a variable such as x or n stand for the unknown number.

An **equation** is a mathematical sentence which states that two expressions are equal.
$$n + 6 = 14$$

Number sentences do not always contain an equal sign. An **inequality** is a mathematical sentence that uses $<$, $>$, $\leq$, or $\geq$ to show the relationship between two expressions.
$$15 + n \geq 26$$

Example: Write an equation or inequality for each word sentence.
1. A number is nine more than twenty. $n = 9 + 20$
2. A number times four minus six is ten. $4n - 6 = 10$
3. A number divided by six is less than two. $n \div 6 < 2$ or $\frac{n}{6} < 2$
4. What number added to 478 gives 1019? $n + 478 = 1019$

Write an equation or inequality for each sentence. Use n for the variable.

1. A number plus one is greater than three.   _____

2. Two less than a number is seven.   _____

3. Ten times a number is fifty.   _____

4. A number is less than twelve.   _____

5. A number and thirty totals sixty-four.   _____

6. A number increased by one is one-half of twenty-four.   _____

7. Seventeen is 42 minus one-half of a number.   _____

8. The product of 12 and a number, divided by 6, is greater than 58.   _____

9. Eighteen times a number is 36.   _____

10. Four more than a number is less than 38.   _____

11. Twenty divided by the difference of 15 and a number is 3.   _____

12. Eight times the sum of a number and 3 is 35.   _____

13. Thirteen and twice a number is less than or equal to 61.   _____

14. Four more than 5 times a number is less than or equal to 39.   _____

Basic Skills Practice

Writing Equations and Inequalities, Part 2

Read each question and circle the best answer.

1. Sixty-eight divided by a number is seventeen. Which equation matches this sentence?

   A $68n = 17$

   B $\dfrac{68}{n} = 17$

   C $68 + 17 = n$

   D $n - 68 = 17$

2. Which equation states that three times a number minus seven is twenty-three?

   F $3n + 7 = 23$

   G $7(3n) = 23$

   H $3n - 7 = 23$

   I $3 + n - 7 = 23$

3. Which sentence is equivalent to $2n + 5 = 15$?

   A Five added to twice a number is fifteen.

   B Twice a number is fifteen.

   C A number plus five is fifteen.

   D Fifteen is a number plus five.

4. A number plus two is greater than or equal to three. Which inequality matches this sentence?

   F $n + 2 > 3$

   G $n \geq 3$

   H $n + 2 \leq 3$

   I $n + 2 \geq 3$

5. Angie's grandfather is three times as old as her brother. If Angie's grandfather is 66, which equation could you use to find Angie's brother's age?

   A $3n = 66$

   B $3 + n = 66$

   C $3 \times 66 = n$

   D $66 - n = 3$

6. Gene swims 3600 meters in 30 minutes. Which equation could be used to find his average speed per minute?

   F $30 \times 3600 = n$

   G $3600 - 30 = n$

   H $\dfrac{3600}{30} = n$

   I $\dfrac{30}{3600} = n$

7. During 1996, the company showed a profit of at least $1500 each month. Which inequality reflects this statement?

   A $n > \$1500$

   B $n \leq \$1500$

   C $n \geq \$1500$

   D $n < \$1500$

8. The longest Viking ship that has been found is a little less than 95 feet long. Which inequality shows the length of the Viking ships found so far?

   F $n + 95$

   G $n > 95$

   H $n < 95$

   I $n \leq 95$

9. On a math test, you could score 100 points and 20 bonus points. Your score on the test was 87, which included 12 bonus points. Which equation shows the score you would receive without the bonus points?

   A $n + 20 = 100$

   B $n + 87 = 100$

   C $n + 20 = 87$

   D $n + 12 = 87$

Basic Skills Practice
Generating Formulas, Part 1

A **formula** is an algebraic sentence. Mathematical formulas are used in science, business, and economics. Letters are chosen to represent various quantities.

To use a formula in solving a problem:
- Choose a formula that fits the problem.
- Replace the variables with given values.
- Compute or solve.

Example 1: In physics, the formula $d = rt$ gives the distance, d, that an object travels in terms of its rate, r, and the time, t, that it travels. What distance would a car travel in 3 hours at a speed of 50 kilometers per hour?

$$\text{Choose} \rightarrow \quad d = rt$$
$$\text{Replace} \rightarrow \quad d = 50 \times 3$$
$$\text{Compute} \rightarrow \quad d = 150 \text{ kilometers}$$

Problem solving also generates formulas, equations, or number sentences.

Example 2: Girl scout cookies cost $6 for a large box and $3 for a small box. Julie sold 5 large boxes and 4 small boxes, Angie sold 3 large boxes and 6 small boxes, and Melissa sold 1 large box and 2 small boxes. Find the total amount received from each girl scout.

l = number of large boxes s = number of small boxes t = total amount
$$(\$6 \times l) + (\$3 \times s) = t$$

Scout	No. large boxes	+	No. small boxes	= Total
Julie	$6 × 5	+	$3 × 4	= $42
Angie	$6 × 3	+	$3 × 6	= $36
Melissa	$6 × 1	+	$3 × 2	= $12

Inequalities may also be used in problem solving.

Example 3: Mr. Rhodes told the student council that in order to have a dance, at least 175 people must attend. The fire laws, however, allow no more than 350 people in the room where the dance will be held. Write an inequality to describe this situation.
$$175 \le x \le 350$$

Solve each problem.

1. The speed limit on Loop 1 in Austin is a maximum of 65 mph and a minimum of 45 mph. Write a number sentence to express this situation.

_____

2. The formula for the area of a rectangle can be found by multiplying the length by the width, $A = lw$. What is the area of a rectangle whose length is 30 meters and whose width is 25 meters?

_____

3. Melissa said that Jean is one-fourth as old as Mr. Adams. If Jean is 9 years old, how old is Mr. Adams?

_____

Basic Skills Practice

Basic Skills Practice

Generating Formulas, Part 2

Read each question and circle the best answer.

1. A cookie recipe calls for $3\frac{1}{2}$ cups of flour. Angie has only a $\frac{1}{2}$-cup measure. How many times must she fill the measuring cup to obtain the required amount of flour?

   A 4

   B $5\frac{1}{2}$

   C 6

   D 7

2. Gene spends $3.50 each day for lunch. Which equation describes the total cost of lunch, t, for any number of days, x?

   F $\frac{\$3.50}{x} = t$

   G $\$3.50(x) = t$

   H $\$3.50 + x = t$

   I $x + t = \$3.50$

3. The cost of a long distance call is $1.75 for the first 3 minutes and $0.10 for each additional minute. Chuck called home and talked for 7 minutes. Which equation would be used to find the cost, c, of the call?

   A $\$1.75 \times 0.10 = c$

   B $\$1.75 + 4 = c$

   C $\$1.75 + (4 \times 0.10) = c$

   D $\$1.75 + 0.10 = c$

4. The Jacksons are planning a trip. Each day they will drive at least 300 miles but not more than 375 miles. Which number sentence best expresses this situation if n represents miles per day?

   F $300 < n < 375$

   G $300 > n > 375$

   H $300 \leq n \leq 375$

   I $300 \geq n \geq 375$

5. The possible outcomes of tossing three coins are shown by the following diagrams. What is the total number of possible outcomes of this experiment?

   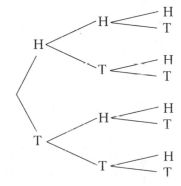

   A 2

   B 4

   C 8

   D 12

The pressure at a certain ocean depth can be represented by the formula $P = 15 + \frac{1}{2}D$, where P stands for the pressure in pounds per square inch and D is the depth in feet. Use this formula to answer each question.

6. What is the ocean depth, in feet, if the pressure is 35 pounds per square inch? _____

7. If the ocean depth is 60 feet, what is the pressure, in pounds per square inch? _____

8. What is the ocean depth, in feet, if the pressure is 50 pounds per square inch? _____

Basic Skills Practice

Cumulative Review

Read each question and circle the best answer.

1. Arthur has 6 times as many stamps as Gene. If Arthur has 108 stamps, which equation would you use to find out how many stamps Gene has?

    A $6 + n = 128$
    B $6n = 108$
    C $6 \times 128 = n$
    D $6 + 108 = n$

2. Which sentence best states that the product of seven and four is greater than nine?

    F $7 \times 4 > 9$
    G $7 + 4 > 9$
    H $\frac{4}{9} > 7$
    I $7 \times 4 < 9$

3. Which set is a true solution for $-3 < n < 5$?

    A $\{-3, -2, -1, 0, 1, 2, 3, 4, 5\}$
    B $\{-3, -2, -1, 0, 1, 2, 3\}$
    C $\{-2, -1, 0, 1, 2, 3, 4\}$
    D $\{-2, -1, 0, 1, 2, 3, 4, 5\}$

4. Which statement is false?

    F $7 + 3 < 11$
    G $-26 \geq 0$
    H $-9 < 0$
    I $8 < 9 \times 2$

5. The maximum grade on a test is 65. The minimum grade is 40. If n represents a grade, which sentence best expresses this situation?

    A $65 \leq n \leq 40$
    B $40 \leq n \geq 65$
    C $65 \geq n \geq 40$
    D $40 \geq n \geq 65$

6. The corner store receives a new delivery of 512 candy bars. Now there are 2115 candy bars in stock. Which equation would you use to determine how many candy bars the store had before the delivery?

    F $2115 + 512 = n$
    G $\frac{2115}{512} = n$
    H $n + 512 = 2115$
    I $512n = 2115$

7. Complete to form an equivalent ratio.
    $$\frac{3}{13} = \frac{9}{x}$$

    A 112
    B 39
    C 27
    D 12

Solve each problem.

8. Pat uses 4 ounces of yeast for every 2 loaves of bread that he bakes. How many pounds of yeast does he need for 40 loaves of bread? _____

9. What number is 4% of 15? _____

Basic Skills Practice

Solving Equations by Addition or Subtraction, Part 1

You can solve an equation by using an inverse operation to get the variable by itself. Add or subtract the same quantity to both sides of the equation to isolate the variable.

Example 1: $n + 3 = 7$ Check: $n + 3 \overset{?}{=} 7$

$n + 3 - 3 = 7 - 3$ $4 + 3 \overset{?}{=} 7$

$n = 4$ $7 = 7$

Example 2: $n - 5 = 12$ Check: $n - 5 \overset{?}{=} 12$

$n - 5 + 5 = 12 + 5$ $17 - 5 \overset{?}{=} 12$

$n = 17$ $12 = 12$

When solving word problems: • Define a variable.
  • Write an equation.
  • Solve and check.

Example 3: Sixty-one points are needed to win a game. Julie has 47 points. How many more points does she need to win?

Define → Let n represent the points needed to win.

Write → $n + 47 = 61$

Solve → $n + 47 - 47 = 61 - 47$

$n = 14$

Check → $n + 47 = 61$

$14 + 47 \overset{?}{=} 61$

$61 = 61$

Julie needs 14 more points.

Solve each equation for *n* and check the solution.

1. $n + 8 = 17$ _____
2. $n + 6 = 15$ _____
3. $n + 16 = 73$ _____
4. $n + 27 = 109$ _____
5. $n + 391 = 487$ _____
6. $n + 195 = 1178$ _____
7. $n - 8 = 36$ _____
8. $n - 6 = 24$ _____
9. $n - 13 = 43$ _____
10. $n - 37 = 35$ _____
11. $n - 190 = 234$ _____
12. $n - 406 = 292$ _____
13. $n + 484 = 581$ _____
14. $n - 80 = 74$ _____
15. $n - 50 = 222$ _____
16. $n + 64 = 71$ _____
17. $n + 37 = 2134$ _____
18. $n - 152 = 634$ _____

Basic Skills Practice

Solving Equations by Addition or Subtraction, Part 2

Read each question and circle the best answer.

1. Coach Hamilton's team ran 20 laps fewer than Coach Faris's team. Coach Faris's team ran 45 laps. How many laps did Coach Hamilton's team run?

   A 25
   B 30
   C 40
   D 65

2. Sara won 9 more games of table tennis than her friend, Chuck. Chuck won 15 games. How many games did Sara win?

   F 6
   G 9
   H 15
   I 24

3. What is the solution to the equation?
   $$n + 213 = 1974$$

   A 213
   B 1761
   C 1974
   D 2187

4. Solve the following equation for n.
   $$n - 64 = 76$$

   F 140
   G 76
   H 64
   I 12

5. It takes Tina 16 minutes to jog 2 miles. Her friend Pamela can jog the same distance in 4 minutes less time. Which equation describes this situation?

   A $n + 16 = 2$
   B $4 + 2 = n$
   C $16 - 4 = n$
   D $16 - n = 4$

6. How long does it take Pamela to jog 2 miles?

   F 20
   G 18
   H 12
   I 6

Solve each problem.

7. The odometer in the school van read 794 on Monday. On Friday, it read 1031. How many miles was the van driven that week? _____

8. Sami bought a mountain bike for $329. She used money in a savings account and a paycheck for $138. How much had Sami saved for the mountain bike? _____

9. Acme Construction had $9498.53 in its checking account. After the weekly deposit, the account balance was $17,478.29. How much was the deposit? _____

Basic Skills Practice

Solving Equations by Multiplication or Division, Part 1

You can solve an equation by using the inverse operation to get the variable by itself. Multiply or divide both sides of the equation by the same quantity.

Example 1: $5n = 15$

$$\frac{5n}{5} = \frac{15}{5}$$

$$n = 3$$

Check:

$$5n = 15$$
$$5 \times 3 \overset{?}{=} 15$$
$$15 = 15$$

Example 2: $\frac{n}{5} = 3$

$$\frac{5}{1}\left(\frac{n}{5}\right) = \frac{3}{1}\left(\frac{5}{1}\right)$$

$$n = 15$$

Check:

$$\frac{n}{5} = 3$$
$$\frac{15}{5} \overset{?}{=} 15$$
$$3 = 3$$

When solving word problems:
• Define a variable.
• Write an equation.
• Solve and check.

Example 3: James bought 6 concert tickets for $168. How much did each ticket cost?

Define → Let *n* represent the cost of one ticket.

Write → $6n = 168$

Solve → $\frac{6n}{6} = \frac{168}{6}$

$$n = 28$$

Check → $6 \times 28 \overset{?}{=} 168$

$$168 = 168$$

Each ticket cost $28.

Solve each equation and check the solution.

1. $7n = 56$ _____

2. $9n = 153$ _____

3. $\frac{n}{9} = 15$ _____

4. $\frac{n}{12} = 7$ _____

5. $7n = 217$ _____

6. $57n = 171$ _____

7. $\frac{n}{32} = 18$ _____

8. $\frac{n}{15} = 8$ _____

Answer the following questions in the space provided.

9. During recent road construction, traffic on the freeway was slowed to $\frac{1}{5}$ the normal speed limit. If traffic was moving at about 12 mph, what is the normal speed limit? _____

10. At the end of the semester, Soon Yi typed 96 words per minute. This was 4 times faster than her speed at the beginning of the course. What was her rate at the beginning of the course? _____

Basic Skills Practice

Solving Equations by Multiplication or Division, Part 2

Read each question and circle the best answer.

1. The school will sponsor up to 16 hockey teams. There are 112 students signed up for hockey. How many teams of 8 players can be formed?

   A 8
   B 14
   C 16
   D 18

2. The Hamiltons rent a cabin at the lake for 12 days at $29 a day. They rent a boat for $50 a day. What is the total amount of rent for the cabin?

   F $948
   G $600
   H $348
   I $398

3. Solve the equation $14n = 350$.

   A 14
   B 25
   C 350
   D 4900

4. Solve $\frac{n}{27} = 754$.

   F 20,358
   G 279
   H 27.9
   I 27

5. On a highway, the fastest car is traveling 75 mph. This is 1.5 times as fast as the slowest car. At what speed is the slowest car traveling?

   A 135 mph
   B 90 mph
   C 50 mph
   D 15 mph

6. Eight adult tickets to a movie cost $56. Children's tickets cost $3.50 each. What is the cost of each adult ticket?

   F $3.50
   G $5.75
   H $7.00
   I $10.50

Solve each problem.

7. Three of the toll booths on the Sam Houston Tollway were closed during rush hour this morning. There were at least 17 cars in line at each of the other 4 tollbooths. How many cars were waiting to pay tolls?

   _____

8. If it takes an average of 45 seconds for each car to go through the toll booth, how long will the 17th car wait in line for its turn to go through the toll booth? Give your answer in minutes; remember that there are 60 seconds in a minute.

   _____

9. If the three toll booths stayed closed all day and 4268 cars went through the remaining toll booths, how many cars went through each of the 4 toll booths? Assume an equal number of cars went through each open toll booth.

   _____

NAME _____ CLASS _____ DATE _____

Basic Skills Practice

Solving Two-Step Equations, Part 1

If an equation contains more than one operation, undo the addition or subtraction first. Then undo the multiplication or division. These steps are performed in the reverse order of the order of operations.

Example 1:
$$2n - 7 = 29$$
$$2n - 7 + 7 = 29 + 7$$
$$\frac{2n}{2} = \frac{36}{2}$$
$$n = 18$$

Check:
$$2n - 7 = 29$$
$$2 \times 18 - 7 \overset{?}{=} 29$$
$$36 - 7 \overset{?}{=} 29$$
$$29 = 29$$

Example 2:
$$\frac{n}{2} + 4 = 10$$
$$\frac{n}{2} + 4 - 4 = 10 - 4$$
$$2 \times \frac{n}{2} = 6 \times 2$$
$$n = 12$$

Check:
$$\frac{n}{2} + 4 = 10$$
$$\frac{12}{2} + 4 \overset{?}{=} 10$$
$$6 + 4 \overset{?}{=} 10$$
$$10 = 10$$

Solve and check.

1. $2n + 5 = 9$ _____

2. $4n + 6 = 18$ _____

3. $3n - 2 = 10$ _____

4. $3n - 6 = 24$ _____

5. $\frac{n}{2} + 5 = 10$ _____

6. $\frac{n}{6} + 6 = 8$ _____

7. $\frac{n}{8} + 4 - 6$ _____

8. $10n - 15 = 45$ _____

9. $6n + 11 = 19$ _____

10. $5n + 5 = 40$ _____

Write an equation and solve.

11. Abraham Lincoln's Gettysburg Address delivered in 1863 refers to 1776 as *"Four score and seven years ago."* Write an equation and solve for the number of years a score (*s*) represents.

    Equation: _____ *s:* _____

12. Four times a number plus seven is fifteen. Write an equation and solve.

    Equation: _____ *n:* _____

13. In January, 12 tune-ups were done at Harold's Garage. The number of snow tire changes was 6 less than 4 times the number of tune-ups. How many snow tire changes were made?

    Equation: _____ *x:* _____

Basic Skills Practice

93

Basic Skills Practice

Solving Two-Step Equations, Part 2

Read each question and circle the best answer.

1. Four times a number decreased by 3 equals 25. Find the number.

    A 3

    B 7

    C 22

    D 28

2. There were 220 people, including spectators and players, at a game. If there are 10 times as many spectators as players, which equation would you use to find the number of players?

    F $10n = 220$

    G $n = \frac{220}{10}$

    H $10n + n = 220$

    I $220 + 10n = n$

3. Six less than twice a number is 14. Find the number.

    A 6

    B 10

    C 14

    D 20

4. Melissa bought three pairs of jeans for $144. One pair cost $9 more than each of the other two. What did she pay for each pair?

    F $36, $36, $36

    G $45, $48, $51

    H $45, $45, $54

    I $48, $48, $48

5. Arthur earned $10 more than twice what his sister Susan earned. Together they earned $250. How much did each one earn?

    A $125, $125

    B $140, $110

    C $150, $100

    D $170, $80

6. $2n + 6 = 28$. What is the value of n?

    F 10

    G 11

    H 12

    I 14

Solve each problem.

7. Angela ordered 263 sugar cones from the Amy's Ice Cream on West Sixth Street. The number of sugar cones was 8 more than 3 times the number of waffle cones. How many waffle cones were ordered? _____

8. Maggie makes sundaes and smoothies at Amy's. Last week, she made 96 smoothies. This was 4 fewer than 5 times the number of sundaes she made. How many sundaes did Maggie make last week? _____

9. A chirping cricket can act as a thermometer. Thirty-seven more than one-fourth of the number of chirps in a minute is about equal to the air temperature in degrees Fahrenheit. What is the temperature, in degrees Fahrenheit, if a cricket is chirping at about 40 chirps per minute? _____

NAME _____ CLASS _____ DATE _____

Basic Skills Practice

Cumulative Review

Read each question and circle the best answer.

1. The odometer on Colby's car read 893 on Monday. On Friday it read 1233. How many miles was the car driven that week?

   A 2126
   B 893
   C 340
   D 300

2. June had $565.15 in her checking account. After June made a deposit, the account balance was $783.96. How much money did June deposit?

   F $218.81
   G $490.06
   H $565.15
   I $783.96

3. The length of a rectangle is 5 centimeters more than its width. The perimeter of the rectangle is 70 centimeters. Find its width and length.

   A 25 cm, 10 cm
   B 15 cm, 20 cm
   C 10 cm, 30 cm
   D 30 cm, 40 cm

4. Miss James traveled 250 kilometers in 5 hours. How far did she travel in 1 hour?

   F 75 km
   G 60 km
   H 50 km
   I 40 km

5. Thomas had $15 in the bank. He worked 10 hours. When he put his paycheck in the bank, his balance was $75. How much did he earn per hour when he worked?

   A $4
   B $5
   C $5.25
   D $6

6. A customer buys a gallon of maple syrup for $24.50. Sales tax is 6%. What is the total cost?

   F $24.56
   G $24.60
   H $25.97
   I $26.00

Solve each problem.

7. A manufacturer found that 2% of the socks produced were unsuitable for marketing and had to be rejected. Of 3500 pairs, how many pairs were rejected?

   _____

8. Misha plans to install wall-to-wall carpet. A rectangular hall on the first floor is 8.5 yards long and 3.0 yards wide. What is the area covered by the carpet in square yards?

   _____

9. In a track meet, Quinton scored 9 points. This was $\frac{1}{3}$ of his team's final score. How many points did the team score?

   _____

Basic Skills Practice

Graphing Points on a Line, Part 1

Example 1: Graph the solution to $x + 7 = 11$.
First solve the problem: $x + 7 = 11$.
Think: $x = 11 - 7 = 4$ (This is an inverse operation.) $x = 4$

Graph:

$$\leftarrow\!\!\!+\quad+\quad+\quad+\quad+\quad+\quad\bullet\quad+\quad+\!\!\!\rightarrow$$
$$-8\quad-6\quad-4\quad-2\quad 0\quad 2\quad 4\quad 6\quad 8$$

Example 2: Graph the solution of $x > 2$.
Solve: $x > 2$.
Think: x can be any number greater than 2.

Graph:

$$\leftarrow\!\!\!+\quad+\quad+\quad+\quad+\quad\oplus\text{======}\rightarrow$$
$$-8\quad-6\quad-4\quad-2\quad 0\quad 2\quad 4\quad 6\quad 8$$

The open circle shows that 2 is NOT a solution.

Example 3: Graph the solution of $x \leq -4$.
Solve: $x \leq -4$.
Think: x can be -4 or any number less than -4.

Graph:

$$\leftarrow\text{======}\bullet\quad+\quad+\quad+\quad+\quad+\quad+\!\!\!\rightarrow$$
$$-8\quad-6\quad-4\quad-2\quad 0\quad 2\quad 4\quad 6\quad 8$$

The closed circle shows that -4 is included in the solution.

Graph the following:

1. $x + 5 = 8$

$$\leftarrow\!\!\!+\quad+\quad+\quad+\quad+\quad+\quad+\quad+\quad+\!\!\!\rightarrow$$
$$-8\quad-6\quad-4\quad-2\quad 0\quad 2\quad 4\quad 6\quad 8$$

2. $x - 4 = 2$

$$\leftarrow\!\!\!+\quad+\quad+\quad+\quad+\quad+\quad+\quad+\quad+\!\!\!\rightarrow$$
$$-8\quad-6\quad-4\quad-2\quad 0\quad 2\quad 4\quad 6\quad 8$$

3. $x < 3$

$$\leftarrow\!\!\!+\quad+\quad+\quad+\quad+\quad+\quad+\quad+\quad+\!\!\!\rightarrow$$
$$-8\quad-6\quad-4\quad-2\quad 0\quad 2\quad 4\quad 6\quad 8$$

4. $x > -4$

$$\leftarrow\!\!\!+\quad+\quad+\quad+\quad+\quad+\quad+\quad+\quad+\!\!\!\rightarrow$$
$$-8\quad-6\quad-4\quad-2\quad 0\quad 2\quad 4\quad 6\quad 8$$

5. $x \geq -2$

$$\leftarrow\!\!\!+\quad+\quad+\quad+\quad+\quad+\quad+\quad+\quad+\!\!\!\rightarrow$$
$$-8\quad-6\quad-4\quad-2\quad 0\quad 2\quad 4\quad 6\quad 8$$

6. $x \leq 5$

$$\leftarrow\!\!\!+\quad+\quad+\quad+\quad+\quad+\quad+\quad+\quad+\!\!\!\rightarrow$$
$$-8\quad-6\quad-4\quad-2\quad 0\quad 2\quad 4\quad 6\quad 8$$

7. $x > -1$

$$\leftarrow\!\!\!+\quad+\quad+\quad+\quad+\quad+\quad+\quad+\quad+\!\!\!\rightarrow$$
$$-8\quad-6\quad-4\quad-2\quad 0\quad 2\quad 4\quad 6\quad 8$$

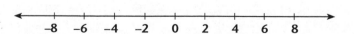

Basic Skills Practice

Graphing Points on a Line, Part 2

Read each question and circle the best answer.

1. Monday's average temperature was 80° F; the temperature varied by 5° F throughout the day. Which number line represents Monday's temperatures?

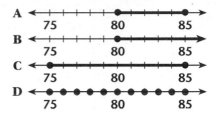

2. The Sanchez family is taking a vacation to Colorado. They want to save at least $750 before their trip for expenses. Which number line shows the amount of money that the Sanchez family would like to save?

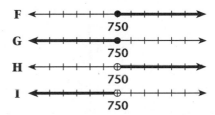

3. Dr. Hughes advised her patient that his weight should not vary more than 5 lb in either direction from 170 lb. Which number line represents this weight range?

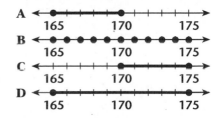

4. The Crunchie Cookie Company asks its truck drivers to drive at least 55 mph but slower than 65 mph. Which number line shows the range of speeds that the company wants its drivers to maintain?

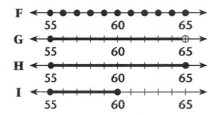

5. Which inequality represents the number line?

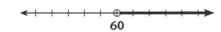

A $x \geq 60$

B $x < 60$

C $x \leq 60$

D $x > 60$

6. Jeremy's mother told him not to spend more than $40 for a new pair of jeans. Which number line shows the price range that Jeremy should stay within?

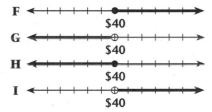

Solve each problem.

7. Shopping for school clothes, Jayna found a jacket that cost $80. The cost of the jacket is $10 more than two pairs of jeans that she also wants to buy. What is the cost of each pair of jeans? _____

8. At a radar checkpoint, 5 cars were traveling at the following speeds: 65 mph, 72 mph, 62 mph, 58 mph, and 70 mph. What was the average speed of the cars, in miles per hour? _____

Basic Skills Practice

Graphing Points on a Coordinate Grid, Part 1

A coordinate grid is made up of the intersection of two lines. The vertical line is called the *y*-axis, and the horizontal line is called the *x*-axis. The lines intersect at the origin. The coordinates of the origin are (0, 0).

Example 1: Write the ordered pair for each point.

Point *A*   From the origin, move right 4 units and down 2 units. The ordered pair is (4, −2).

Point *B*   From the origin, move left 3 units and up 5 units. The ordered pair is (−3, 5).

Example 2: Write the point named by each ordered pair.

(1, 3)   From the origin, move right 1 unit and up 3 units. Point *C* is located at (1, 3).

(−2, −3)   From the origin, move left 2 units and down 3 units. Point *D* is located at (−2, −3).

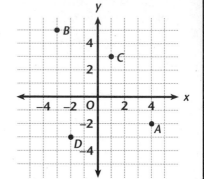

Use the coordinate grid at right for Exercises 1–8. Write the ordered pair.

1. point *A* _____

2. point *B* _____

3. point *C* _____

4. point *D* _____

5. point *E* _____

6. point *F* _____

7. point *G* _____

8. point *H* _____

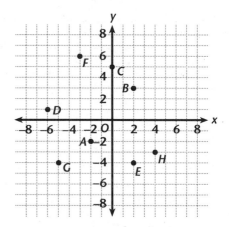

Use the coordinate grid at right for Exercises 9–20. Write the point named by the ordered pair.

9. (6, −7) _____

10. (−3, −6) _____

11. (5, 5) _____

12. (−4, 7) _____

13. (−2, 3) _____

14. (1, 6) _____

15. (−4, −3) _____

16. (3, −2) _____

17. (−2, −5) _____

18. (4, 2) _____

19. (5, −1) _____

20. (−5, 4) _____

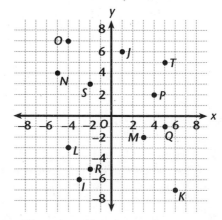

Basic Skills Practice

Graphing Points on a Coordinate Grid, Part 2

Read each question and circle the best answer.

Use the graph to answer Exercises 1 and 2.

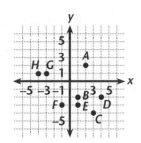

1. For which point is $x > 1$ and $y < -2$?

   A point A

   B point B

   C point C

   D point D

2. For which point is $x < -3$ and $y \geq 1$?

   F point E

   G point F

   H point G

   I point H

3. What is the ordered pair for point A?

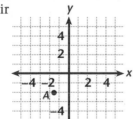

   A $(-1.5, -2)$

   B $(-2, -2)$

   C $(-2, -1.5)$

   D $(1.5, 2)$

4. Which is an equation for line m?

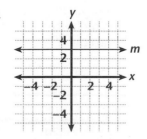

   F $x = 3$

   G $y = 3$

   H $y = -3$

   I $x = -3$

Use the graph to answer Exercises 5 and 6.

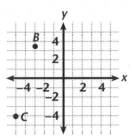

5. What are the coordinates of point B?

   A $(-3, 4)$

   B $\left(3, -3\frac{1}{2}\right)$

   C $\left(-3, 3\frac{1}{2}\right)$

   D $\left(-3\frac{1}{2}, 3\right)$

6. What are the coordinates of point C?

   F $(-5, -4)$

   G $(-5, 4)$

   H $(5, -4)$

   I $(-4, -5)$

7. For which point is $x > -1$ and $y \leq -2$?

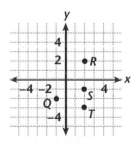

   A point Q

   B point R

   C point S

   D point T

8. What is the ordered pair for point S?

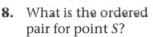

   F $(4.5, 3)$

   G $(-4.5, 3)$

   H $(3, -4.5)$

   I $(4.5, -3)$

Basic Skills Practice

Cumulative Review

Read each question and circle the best answer.

1. If $4m - 5 = -29$, what is the value of m?

   A -6
   B -1
   C 6
   D 24

2. On a farm-to-market road the maximum speed limit is 55 miles per hour, and the minimum speed is 35 miles per hour. If x represents the speed, which sentence best describes this condition?

   F $35 \le x \ge 55$
   G $35 \ge x \ge 55$
   H $55 \le x \le 35$
   I $55 \ge x \ge 35$

3. The cost of placing an ad in the local paper is given by the formula $C = \$4.00 + \$0.45w$, where C is the total cost and w is the number of words. What is the cost of an ad containing 22 words?

   A $\$4.45$
   B $\$9.90$
   C $\$13.90$
   D $\$19.80$

4. If $3.4n + 8 = 31.8$, what is the value of n?

   F 2
   G 4
   H 7
   I 8

5. The average weight of a five-year-old child is 40 pounds. Most children's weight varies by 4 pounds. Which number line shows the weight range for a five-year-old?

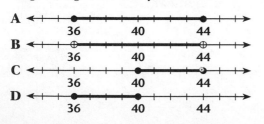

Use the graph to answer Exercises 6 and 7.

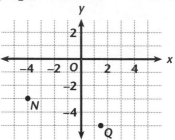

6. What is the ordered pair for point N?

   F $(4, -3)$
   G $(-4, -3)$
   H $(4, 3)$
   I $(-3, -4)$

7. What is the ordered pair for point Q?

   A $(1, 5)$
   B $(-1.5, -5)$
   C $(1.5, -5)$
   D $(1.5, 5)$

8. Which inequality or equation represents the number line below?

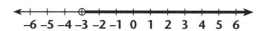

   F $x < -3$
   G $x \ge -3$
   H $x = -3$
   I $x > -3$

9. The average temperature in Austin, Texas, in the month of July is 85 degrees. This temperature often varies by 15 degrees. Which number line below shows this variance?

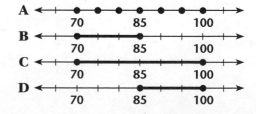

Basic Skills Practice

Finding Perimeter and Using the Distance Formula, Part 1

To find the perimeter, add each length to find the total distance around the outside of a figure.

Example 1: Find the perimeter of the rectangle shown.
Add the length of all sides to get the total.
$P = 3$ ft $+ 4$ ft $+ 3$ ft $+ 4$ ft
$P = 14$ ft

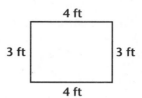

To find the distance that an object travels, multiply its rate of speed by the length of time that it travels.

Example 2: How far will a truck go in 4 hours at 60 mph?
$d = rt$ → Write the formula ($d =$ distance, $r =$ rate, $t =$ time).
$d = 60(4)$ → Substitute the given values ($r = 60$, $t = 4$).
$d = 240$ → Multiply.
The truck will go 240 miles.

Find the perimeter.

1.
5 ft
5 ft

_____

2.
10 yd
4 yd
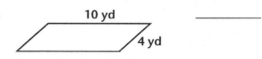
_____

3.
4 m
2 m
5 m
3 m
7 m
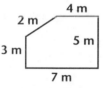
_____

4.
8 in.
4 in.
7 in.
_____

5.
3 cm   3 cm
3 cm   3 cm
3 cm   3 cm
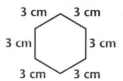
_____

6.
7 mm   7 mm
2 mm
_____

Find the missing amount. If necessary round the answers to the nearest hundredth.

7. $r = 50$ mph, $t = 6$ h, $d =$ _____ mi

8. $r = 32$ mph, $t = 4.5$ h, $d =$ _____ mi

9. $r = 475$ mph, $t =$ _____ h, $d = 2150$ mi

10. $r =$ _____ mph, $t = 1.5$ h, $d = 70.25$ mi

11. $r = 66$ mph, $t = 0.8$ h, $d =$ _____ mi

12. $r = 105$ mph, $t =$ _____ h, $d = 43$ mi

Basic Skills Practice

Finding Perimeter and Using the Distance Formula, Part 2

Read each question and circle the best answer.

1. How many yards of fence would it take to enclose a garden with a width of 10 yards and a length of 12 yards?

   A 42 yd

   B 38 yd

   C 44 yd

   D 22 yd

2. A cargo plane flew 100 miles at 200 mph. How long did the flight take?

   F 2 hr

   G 1 hr

   H 30 min

   I 1 hr 30 min

3. A cowboy built a square pen for his cattle. Each side is 50 feet long. What is the perimeter?

   A 200 ft

   B 500 ft

   C 250 ft

   D 2500 ft

4. Each side of a certain hexagon is 6 inches long. What is the perimeter of the hexagon?

   F 60 in.

   G 36 in.

   H 38 in.

   I 40 in.

5. A bullet-train travels 300 miles in 2 hours. What is the average speed of the train?

   A 100 mph

   B 250 mph

   C 200 mph

   D 150 mph

6. Louis rides his bicycle for 10 minutes at an average speed of 30 mph. How far did he ride?

   F 3 mi

   G 5 mi

   H 6 mi

   I 8 mi

Solve each problem. If necessary round the answer to the nearest hundredth.

7. A football field is 30 yards wide and 100 yards long. What is the perimeter in yards? _____

8. A certain county is perfectly square. Each side is 40 miles long. What is the perimeter in miles? _____

9. An F-16 jet flew 5127.5 miles at an average speed of 1520 mph. How long did the trip take in hours? _____

10. A UFO travels at a rate of 45,500 mph. How long does it take to travel 967,350 miles in hours? _____

11. The length of a basketball court in the school gymnasium measures 28 meters and the width measures 50 meters. Find the perimeter in meters. _____

Basic Skills Practice

Areas of Triangles, Rectangles, and Squares, Part 1

Area is the number of square units needed to cover the surface of a plane figure.

Example 1: To find the area of a rectangle multiply the length by the width.
Using the formula, $A = l \times w$, let $l = 4$ ft and $w = 3$ ft.
$$A = 4 \text{ ft} \times 3 \text{ ft}$$
$$A = 12 \text{ ft}^2$$

Example 2: To find the area of a square, square the length of one side.
Using the formula, $A = s^2$, let $s = 3$ m.
$$A = 3 \text{ m} \times 3 \text{ m}$$
$$A = 9 \text{ m}^2$$

Example 3: To find the area of a triangle multiply one-half of the base times the height.
Using the formula, $A = \frac{1}{2}bh$, let $b = 3$ ft and $h = 4$ ft.
$$A = \frac{1}{2}(3 \text{ ft})(4 \text{ ft})$$
$$A = 6 \text{ ft}^2$$

Find the area of each rectangle.

1. $5\frac{1}{2}$ in. $3\frac{1}{4}$ in. _____

2. 5.2 cm 2.7 cm _____
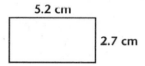

3. $l = 9\frac{1}{4}$ yd, $w = 3\frac{1}{2}$ yd _____

4. $l = 4.4$ mm, $w = 2.9$ mm _____

Find the area of each square.

5. $2\frac{1}{2}$ in. _____

6. 18 mm _____

7. $s = 7.8$ cm _____

8. $s = 16$ in. _____

Find the area of each triangle.

9.   3.2 mm 6.8 mm _____

10.  10 mm 28 mm _____

11. $b = 11$ cm, $h = 23$ cm _____
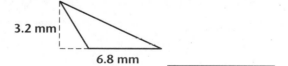

12. $b = 4.5$ in., $h = 6.5$ in. _____
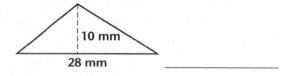

Basic Skills Practice

Areas of Triangles, Rectangles, and Squares, Part 2

Read each question and circle the best answer.

1. John is building an additional room that is 15 feet by 24 feet. How many *square yards* of carpet are needed to cover the floor?

   A 39 yd²
   B 40 yd²
   C 45 yd²
   D 360 yd²

2. The frame around a picture has a length of 20 inches and a width of 16 inches. The length of the picture inside the frame is 14 inches and the width is 10 inches. What is the area of the frame?

   F 180 in.²
   G 320 in.²
   H 220 in.²
   I 460 in.²

3. Alex plans to install wall-to-wall carpeting in his house. He needs to buy 156.25 square yards to cover a square room. What are the dimensions of the room?

   A 12.5 yd × 12.5 yd
   B 15.5 yd × 15.5 yd
   C 17.5 yd × 17.5 yd
   D 20.5 yd × 20.5 yd

4. What is the area of a rectangular house with a length of 30 feet and a width of 20 feet?

   F 600 ft²
   G 500 ft²
   H 50 ft²
   I 1100 ft²

Solve each problem.

5. Keisha plans to buy flowers to cover a triangular garden in her backyard. The dimensions of the garden are 12 feet along the bottom and 16 feet along one side. A 20 foot side completes the garden and forms a right triangle. What is the area of the garden in square feet?

6. Eric plans to install wall-to-wall carpeting in his square study. The floor is 5.5 yards wide. What is the area to be covered by the carpet in square yards? _____

7. In the same house, Eric carpets a rectangular room that is 7.7 yards long and 6.2 yards wide. How much more area, in square yards, is covered by this carpet than by the carpet in Exercise 6? _____

8. Calvin plans to paint the four walls and the ceiling in a guest room. Each wall is rectangular with a width of 12 feet and a height of 10 feet. The ceiling is a square with sides of 12 feet. Find the total area, in square feet, of the ceiling and the four walls. _____

9. The area of a rectangular field is 165,000 square meters. If the length is about 480 meters, what is the approximate width, in meters, of the field? _____

Basic Skills Practice

Cumulative Review

Read each question and circle the best answer.

1. Which integer is equivalent to the following expression:
$$5 \times 3^2 - 10$$

   A −45

   B −5

   C 35

   D 215

2. How many square feet of sod are needed to cover a 40-foot-by-65-foot lawn?

   F 105 ft²

   G 210 ft²

   H 2400 ft²

   I 2600 ft²

3. The distance around a square swimming pool is 48 feet. What are the dimensions of the pool?

   A 12 ft by 12 ft

   B 14 ft by 14 ft

   C 16 ft by 16 ft

   D 24 ft by 24 ft

4. Calculate the following expression:
$$12 - 4.031$$

   F 4.019

   G 4.021

   H 7.969

   I 8.031

5. Roy bought carpet for two bedrooms. One room is 9.5 ft by 11.5 ft and the other is 14 ft by 15 ft. About how many square feet of carpeting did he buy?

   A 50 ft²

   B 60 ft²

   C 300 ft²

   D 319 ft²

6. Joe saved $17.80 on a jacket he bought on sale. The regular price for the jacket was $89.00. What percent of the regular price did he save?

   F 10%

   G 20%

   H 25%

   I 30%

Solve each problem.

7. Lauren walked around her office building during her lunch hour. She walked 0.2 miles along one side of the building and 0.45 miles along another. She cut through the courtyard, which was 0.3 miles, and came back to her desk, which was another 0.25 miles. What was the total distance she walked in miles? _____

8. Brandon is preparing to plant a garden. He wants it to be 6 feet along one side and 8 feet along another. Then he wants these two sides to be connected by a 10-foot side. What is the area of the garden in square feet if the garden is a right triangle? _____

Basic Skills Practice

Circumference and Area of Circles, Part 1

Circumference is the distance around a circle. To find the circumference, multiply the length of the diameter by π. Diameter is the distance across the circle and π is approximately 3.14.

Example 1: Find the circumference of the circle.

$C = \pi d$ →   Write the formula.
$C \approx 3.14(5)$ →   Substitute given values.
$C \approx 15.7$ →   Multiply.

To find the area of a circle, multiply π by the square of the radius (r^2). Remember that the radius is half the diameter of a circle.

Example 2: Find the area of the circle.

$A = \pi r^2$ →   Write the formula.
$A \approx 3.14(3)(3)$ →   Substitute given values.
$A \approx 28.26$ →   Multiply.

Find the circumference for each circle.

1.

_____

2.

_____

3.

_____

4.

_____

Find the area for each circle.

5.

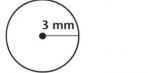

_____

6.

_____

7.

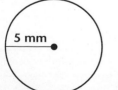

_____

8.

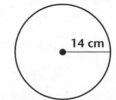

_____

Basic Skills Practice

Circumference and Area of Circles, Part 2

Read each question and circle the best answer.

1. A circular track is 300 yards across. What is its circumference?

   A 390 yd
   B 938 yd
   C 942 yd
   D 953 yd

2. A circular cooking pan has a radius of 6 inches. What is its area in square inches?

   F 94.4 in.^{2}
   G 113.04 in.^{2}
   H 125.01 in.^{2}
   I 154.38 in.^{2}

3. The face of a clock has a diameter of 14 inches. What is its area?

   A 153.86 in^2
   B 172.81 in^2
   C 189.2 in^2
   D 205.65 in^2

4. A certain castle has a circular tower. The circumference of the tower is 208 feet. What is the diameter of the tower?

   F 29.2 ft
   G 43.81 ft
   H 58.3 ft
   I 66.24 ft

5. The diameter of a large coin is 1.5 inches. What is its area in square inches?

   A 1.77 in.^{2}
   B 2.86 in.^{2}
   C 3.07 in.^{2}
   D 4.22 in.^{2}

6. The radius of a certain spherical asteroid is 6 kilometers. What is its circumference?

   F 20.56 km
   G 27.32 km
   H 32.29 km
   I 37.68 km

Solve each problem. If necessary, round answers to the nearest hundredth.

7. A round sign has a diameter of 22.5 feet. What is its area in square feet? _____

8. Another sign has a diameter of 21.5 inches. What is its area in square inches? _____

9. The base of a coffee mug has a diameter of 3.2 inches. What is its circumference in inches? _____

10. The base of a paper cup has a diameter of 2.8 inches. What is its area in square inches? _____

11. Find the circumference, in inches, of a circle that has a diameter of 9 inches. _____

12. Find the area, in square inches, of a circle that has a diameter of 44 inches. _____

13. Find the circumference, in inches, of a basketball with a diameter of 11 inches. _____

Basic Skills Practice

Surface Area and Volume, Part 1

The amount of material needed to cover an object is known as its surface area.
Volume is the number of nonoverlapping unit cubes that will fill the interior of a solid.

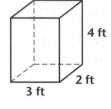

Example 1: Find the area and volume of the rectangular box.
- Find the area of each face.
  Use $A = l \times w$.
  Top and bottom: $3 \times 2 = 6$ ft² each
  Front and back: $3 \times 4 = 12$ ft² each
  Sides: $4 \times 2 = 8$ ft² each
- Add the area of each face.
  $SA = (2 \times 6 \text{ ft}^2) + (2 \times 12 \text{ ft}^2) + (2 \times 8 \text{ ft}^2) = 52 \text{ ft}^2$
- Use the formula $V = lwh$ to find the volume.
  $V = (2 \text{ ft})(3 \text{ ft})(4 \text{ ft}) = 24 \text{ ft}^3$

Example 2: Find the surface area and volume of the cube.
- Find the area of one face.
  Use $A = s^2$.
- Since all the faces are the same, multiply
  the area of one face by the total number of
  faces (6).
  $SA = 6(25 \text{ in.}^2) = 150 \text{ in.}^2$
- Use the formula $V = s^3$ to find the volume.
  $V = (5 \text{ in.})^3 = 125 \text{ in.}^3$

Example 3: Find the lateral surface area (curved part without ends)
and the volume of the cylinder.
- Use $SL = 2\pi rh$.
  $SL = 2\pi(3 \text{ cm})(8 \text{ cm}) = 2\pi(24) \text{ cm}^2 = 48\pi \text{ cm}^2$
- If it is necessary to eliminate π, then use $\frac{22}{7}$.
  $SL = 48\left(\frac{22}{7}\right) \text{ cm}^2 \approx 151 \text{ cm}^2$
- Use the formula $V = \pi r^2 h$ to find the volume.
  $V = \pi(3 \text{ cm})^2(8 \text{ cm}) = \pi(9 \text{ cm}^2)(8 \text{ cm}) = 72\pi \text{ cm}^3$ or 226 cm^3

Find the surface area and volume of each solid figure. If necessary round the answers to the nearest tenth.

1.
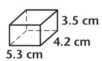
3.5 cm
4.2 cm
5.3 cm

Surface area: _____ Volume: _____

2.

2 m 5 m
1 m

Surface area: _____ Volume: _____

3.

4.2 cm
4.2 cm
4.2 cm

Surface area: _____ Volume: _____

4.

5 in.
2 in.

Lateral SA: _____ Volume: _____

Basic Skills Practice

Surface Area and Volume, Part 2

Read each question and circle the best answer.

1. A can of baked beans requires a label to cover the curved part of the can without overlapping. The radius of the can is 3 inches. The height of the can is 8 inches. Find the size of the label.

   A 12π in.^{2}

   B 24π in.^{2}

   C 48π in.^{2}

   D 72π in.^{2}

2. Al needs to paint the surfaces of a rectangular toy box. The box measures 3 feet in length and 2 feet in width. If 1 pint of paint covers 25 square feet, what else does Al need to know in order to find the number of pints that he needs?

   F the weight of the box

   G the number of ounces in 1 pint

   H the cost of 1 ounce of paint

   I the height of the box

3. James wants to know how much water his swimming pool will hold. The pool is 5 meters by 10 meters by 3 meters. How much water will the pool hold?

   A 50 m^3

   B 150 m^3

   C 750 m^3

   D 1000 m^3

4. Sara wants to cover a cube with colorful contact paper. She can cut the paper the exact size of each side of the cube. Then she can stick the paper onto each side. Find out how much contact paper she will need if one side of the cube is $1\frac{1}{2}$ ft.

   F 2.25 ft^2

   G 10.13 ft^2

   H 13.5 ft^2

   I 18 ft^2

5. Jose wants to mail a gift to his mother. The box he wants to mail it in is the shape of a cube. The cost to mail the gift depends on the amount of space it occupies. What information will be the most helpful to determine the cost to mail the gift?

   A the volume of the box

   B the surface area of the box

   C the weight of the box

   D the shape of the box

6. What is the volume of a cube-shaped fish tank whose side is equal to 2.5 feet?

   F 6.25 m^3

   G 7.5 m^3

   H 15 m^3

   I 15.625 m^3

Solve each problem.

7. A box holds 12 pieces of chalk. The radius of each piece of chalk is 2 centimeters and the height of each piece of chalk is 7 centimeters. What is the total volume of each piece of chalk? Use $\pi = \frac{22}{7}$. _____

8. Smith Wood Supply makes crates of many sizes and shapes. One standard crate measures 7 feet by 4 feet by 2 feet. What number would best represent the amount of gravel the crate can hold? _____

Basic Skills Practice

Cumulative Review

Read each question and circle the best answer.

1. The cost of having a birthday party at a pizza restaurant is given by the formula

    $$C = \$20.00 + \$2.50p,$$

    where p is the number of people coming to the party. What is the cost of a pizza party with 15 people attending?

    A $22.50

    B $37.50

    C $57.50

    D $42.50

2. A crayon company wants to cover each crayon with paper that does not overlap. The radius of each crayon is 3 centimeters and the height is 8 centimeters. Find the amount of paper that it will take to cover each crayon.

    F 12π cm^2

    G 24π cm^2

    H 48π cm^2

    I 72π cm^2

3. 9.067 is between which pair of consecutive integers?

    A 8 and 9

    B 9 and 10

    C 10 and 11

    D 30 and 31

4. The diameter of a basketball hoop is 55 centimeters. A manufacturer wants to know how much metal it will take to make the rim. Find the circumference of the rim. Use 3.14 for π.

    F 55 cm

    G 86.4 cm

    H 172.7 cm

    I 345.4 cm

5. Refer to the information given in Exercise 4. About how much space within the basketball hoop will there be for a basketball player to make a basket?

    A 55π cm^2

    B 110π cm^2

    C 756π cm^2

    D 3025π cm^2

6. Elizabeth enjoys wearing ribbon in her hair. She has 6 red, 4 green-and-white striped, 2 blue, and 8 floral ribbons. What fraction of the total number of ribbons are green-and-white striped?

    F $\dfrac{1}{5}$

    G $\dfrac{1}{10}$

    H $\dfrac{3}{10}$

    I $\dfrac{2}{5}$

Solve each problem.

7. Michele gets paid $7.50 for 2 hours of baby-sitting. Friday night she baby-sat for the Andersons. They paid her $18.75. How many hours did she baby-sit? _____

8. Unleaded gasoline costs $1.259 per gallon. You need 10.4 gallons to fill your gas tank. Rounded to the nearest hundredth, about how much does it cost to fill the tank? _____

110   **Basic Skills Practice**

Basic Skills Practice

Similar Figures, Part 1

Two or more figures that have the same shape but different sizes are called *similar figures.*
Triangle *ABC* and triangle *DEF* are similar figures.
This is written as △*ABC* ~ △*DEF*.

When two figures are similar,
- the corresponding angles are congruent.
  $\angle A \cong \angle D$ $\angle B \cong \angle E$ $\angle C \cong \angle F$

- the ratios of the lengths of the corresponding sides are equal.
  $\dfrac{AB}{DE} = \dfrac{5}{10} = \dfrac{1}{2}$ $\dfrac{BC}{EF} = \dfrac{4}{8} = \dfrac{1}{2}$ $\dfrac{AC}{DF} = \dfrac{3}{6} = \dfrac{1}{2}$

You can determine if two figures are similar by checking to see that they meet both conditions.
Example 1: Are rectangle *ABCD* and rectangle *EFGH* similar?

- Because both figures are rectangles, all angles are right angles, so they are all
  congruent.
- $\dfrac{AB}{EF} = \dfrac{2}{3}$ $\dfrac{BC}{FG} = \dfrac{4}{6} = \dfrac{2}{3}$ $\dfrac{CD}{GH} = \dfrac{2}{3}$ $\dfrac{DA}{HE} = \dfrac{4}{6} = \dfrac{2}{3}$

The ratios of the corresponding sides are equal, so rectangle *ABCD* ~ rectangle
EFGH.

You can find the length of the missing side in similar figures by using proportions.

Example 2: The two flags are similar. What is the length of *s*?

Write a proportion.   → $\dfrac{2}{3} = \dfrac{4}{s}$

Solve the proportion.   → $2s = 12$
                              $s = 6$

**Determine if the pairs of figures are similar. Write *yes* or *no* in
the blank spaces.**

1.   _____

2.    _____

Each pair of figures is similar. Find the length of the missing side.

3. 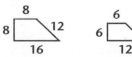   _____

4.    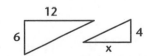 _____

Basic Skills Practice

Similar Figures, Part 2

Read each question and circle the best answer.

In Exercises 1–4, each pair of figures is similar. Find the length of the missing side.

1.

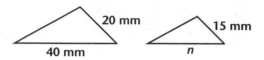

20 mm
40 mm
15 mm
n

    A   *n* = 20 mm

    B   *n* = 25 mm

    C   *n* = 30 mm

    D   *n* = 35 mm

2.

20 mm
25 mm
n
30 mm

    F   *n* = 22 mm

    G   *n* = 24 mm

    H   *n* = 26 mm

    I   *n* = 28 mm

3.

n
3 cm
2 cm
4 cm

    A   *n* = 1.0 cm

    B   *n* = 1.5 cm

    C   *n* = 2.0 cm

    D   *n* = 2.5 cm

4.

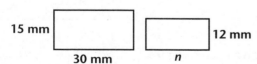

15 mm
30 mm
12 mm
n

    F   *n* = 26 mm

    G   *n* = 24 mm

    H   *n* = 22 mm

    I   *n* = 20 mm

In Exercises 5–9, $\triangle ABC \sim \triangle DEF$.

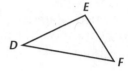

5. $\angle A \cong$ _____

    A $\angle B$

    B $\angle D$

    C $\angle E$

    D $\angle F$

6. $\angle B \cong$ _____

    F $\angle F$

    G $\angle E$

    H $\angle D$

    I $\angle B$

7. $\dfrac{AB}{DE} = \dfrac{AC}{n}$

    A   *n* = EF

    B   *n* = DF

    C   *n* = DE

    D   *n* = BC

8.   Which side corresponds to $\overline{BC}$?

    F $\overline{BA}$

    G $\overline{CA}$

    H $\overline{EF}$

    I $\overline{DF}$

9.   If $m\angle F = 35°$, what is $m\angle C$?

    A   25°

    B   30°

    C   33°

    D   35°

Basic Skills Practice

Basic Skills Practice

Symmetry and Rotations, Part 1

A figure has *line symmetry* if it can be folded through a line and both parts of the figure coincide or match one another. Since the parts form mirror images of one another, this is also called reflectional symmetry.

Figures may have one line of symmetry, two or more lines of symmetry, or no line of symmetry.

Example 1: Find the lines of symmetry for each figure.

a.   b.   c.

   a. This figure has a horizontal line of symmetry.
   b. This figure has a vertical and a horizontal line of symmetry.
   c. This figure has two congruent parts, but it has no line of symmetry.

Rotational symmetry is another kind of symmetry. A figure has rotational symmetry if there is a point *O* about which the figure can be rotated and made to coincide with its original figure. Rotational symmetries are identified by the angle of rotation required to make the figure coincide with itself. If a figure coincides with itself after being rotated 180°, the figure has *point symmetry*. If a figure only has 360° rotational symmetry, it is not considered to be symmetric because any figure will coincide with itself after being rotated in a full circle.

Example 2: Determine if each figure has rotational symmetry.

a.   b.   c.

   a. This figure has rotational symmetry. It will coincide with itself after being rotated 120°.
   b. This figure does not have rotational symmetry. (It does have vertical line symmetry.)
   c. This figure has point symmetry. It will coincide with itself after being rotated 180°.

Identify the type of symmetry, if any, in each figure.

1.   2.   3.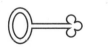

_____   _____   _____

4.   5.   6.

_____   _____   _____

NAME _____ CLASS _____ DATE _____

Basic Skills Practice

Symmetry and Rotations, Part 2

Read each question and circle the best answer.

1.

A Horizontal line of symmetry
B Vertical and horizontal lines of symmetry
C Point symmetry
D No symmetry

2.

F Vertical line of symmetry
G Horizontal line of symmetry
H Rotational symmetry
I No symmetry

3.

A Point symmetry
B Rotation symmetry
C Horizontal line of symmetry
D All of the above

4.

F No symmetry
G Rotational symmetry
H Horizontal line of symmetry
I Vertical line of symmetry

5.

A No symmetry
B Point symmetry
C Horizontal line of symmetry
D Vertical line of symmetry

6.

F Vertical line of symmetry
G Rotational symmetry
H Horizontal line of symmetry
I All of the above

7.

A No symmetry
B Rotational symmetry
C Horizontal line of symmetry
D Vertical line of symmetry

8.

F Vertical line of symmetry
G Horizontal line of symmetry
H Vertical and horizontal lines of symmetry
I No symmetry

114

Basic Skills Practice

Cumulative Review

Read each question and circle the best answer.

1. Write a mathematical sentence for this statement: 8 subtracted from 3 times a number is 16.

   A $8 - 3n = 16$

   B $3n - 8 = 16$

   C $16 - 3n = 8$

   D $16 - 8 = 3n$

2. Find the perimeter of a rectangle with sides of 4 cm and 6 cm.

   F 10 cm

   G 15 cm

   H 20 cm

   I 24 cm

3. Given the similar figures shown, find the length of side *AB*.

   A 4 cm

   B 6 cm

   C 8 cm

   D 10 cm

4. Identify the type of symmetry for the figure shown.

   F Horizontal line of symmetry

   G Vertical line of symmetry

   H Point symmetry

   I No symmetry

5. The formula for the area of a triangle is $\frac{1}{2}bh$, where *b* is the base and *h* is the height. Find the area of a triangle with a base of 5 cm and a height of 3 cm.

   A 15 cm^2

   B 7.5 cm^2

   C 5 cm^2

   D 3.5 cm^2

6. Which inequality describes the number line?

   F $x \le 3$

   G $x \ge -2$

   H $x < 6$

   I $x < 0$

7. What is the volume of a cargo box with a base that measures 20 ft^2 and a height of 5 ft?

   A 25 ft^3

   B 50 ft^3

   C 100 ft^3

   D 125 ft^3

8. To solve the equation $3a + 2 = 11$, you need to—

   F add 2 to both sides, and then divide both sides by 3

   G subtract 2 from both sides, and then divide both sides by 3

   H divide both sides by 3

   I divide both sides by 11

Basic Skills Practice

Reflections and Translations, Part 1

A *reflection* is a special type of transformation that involves flipping a figure over a line called the line of reflection. Line *l* is a line of reflection. The original figure, $\triangle ABC$, is called the pre-image. The figure resulting from the transformation, $\triangle DEF$, is called the image. Pre-images and images have the same size and shape, and are always congruent.

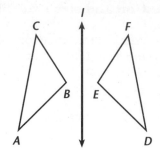

A *translation* is a transformation that moves all the points in a figure a fixed distance in a given direction. A translation retains the same size and shape of the original figure, so the original figure and the translation image are always congruent. Translations are also called *slides*.

Example: Translate $\triangle ABC$ 6 units to the right and 1 unit up, and label the translation image $\triangle XYZ$.

$A(1, 1) \rightarrow X(1 + 6, 1 + 1)$ or $X(7, 2)$
$B(6, 1) \rightarrow Y(6 + 6, 1 + 1)$ or $Y(12, 2)$
$C(1, 5) \rightarrow Z(1 + 6, 5 + 1)$ or $Z(7, 6)$

The graph shows that $\triangle XYZ$ is congruent to $\triangle ABC$.

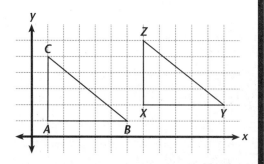

The *y*-axis is the line of reflection in the diagram shown.

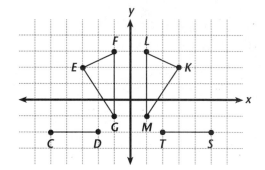

1. Give the coordinates of the reflection of point *E*. _____

2. What is the reflected image of $\triangle EFG$? _____

3. Give the coordinates of the image point for *C*. _____

4. What is the reflected image of segment *CD*? _____

Find the translation image of each of the figures whose vertices are provided.

5. $A(2, 2)$, $B(7, 2)$, $C(7, 5)$, $D(2, 5)$; slide 2 units right and 3 units up

   _____

6. $A(-3, 2)$, $B(-8, 2)$, $C(-8, 4)$, $D(3, 4)$; slide 2 units left and 1 unit up

   _____

7. $A(3, -2)$, $B(6, 2)$, $C(6, -4)$, $D(3, -4)$; slide 3 units right and 2 units down

   _____

8. $A(-3, 1)$, $B(-1, 3)$, $C(0, 0)$; slide 2 units right and 1 unit down

   _____

Basic Skills Practice

Basic Skills Practice

Reflections and Translations, Part 2

Read each question and circle the best answer.

1. In the figure, *Y* is the reflected image of *X*. What is the measure of the angle formed by the line of reflection and the segment connecting points *X* and *Y*?

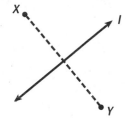

A 55°

B 60°

C 90°

D 180°

2. A line segment with endpoints at (2, 2) and (2, −2) is reflected over the *y*-axis. The endpoints of the reflected image are—

F (−2, 2) and (−2, −2)

G (0, 2) and (0, −2)

H (−1, 2) and (−1, −2)

I (0, −1) and (0, 1)

3. The drawing shows an example of—

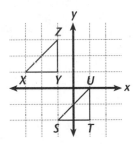

A a translation

B a reflection

C a rotation

D an inflection

4. Suppose △*ABC* is translated 3 units down and 1 unit to the left to create △*DEF*. The vertices of △*ABC* are (1, 2), (4, 3), and (3, 1), respectively. What are the coordinates of the vertices of the image △*DEF*?

F *D*(1, 2), *E*(4, 3), *F*(3, 1)

G *D*(0, 1), *E*(2, 1), *F*(3, 1)

H *D*(0, −1), *E*(3, 0), *F*(2, −2)

I *D*(2, −2), *E*(3, 0), *F*(0, −1)

5. Rectangle *ABCD* is translated, forming image rectangle *FGHI*. What best describes the relationship between rectangle *ABCD* and rectangle *FGHI*?

A They are similar.

B They are congruent.

C They are mirror images.

D They are dilations.

6. A line segment with endpoints at (3, 1) and (2, 4) is reflected over the *y*-axis. The endpoints of the reflected image are—

F (−3, 1) and (−2, 4)

G (1, 3) and (2, −4)

H (−1, 3) and (−2, −4)

I (3, −1) and (2, 4)

7. The drawing shows an example of—

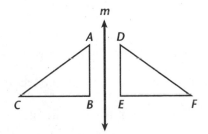

A a translation

B a reflection

C a rotation

D an inflection

Basic Skills Practice

Scale and Dilations, Part 1

A dilation is a transformation that changes the size of a figure while preserving the orientation, shape, and measures of the angles in the figure. Dilations refer to both enlargements and reductions in the size of the original figure. A scale factor or multiplier determines the amount of enlargement or reduction of the original figure. A dilation is a reduction if the scale factor is less than 1, and it is an enlargement if it is greater than 1.

Example 1: Transform segment AB using its endpoints $A(4, 0)$ and $B(2, 3)$ and a scale factor of $\frac{1}{2}$.

Multiply both coordinates of each point.

$A(4, 0) \rightarrow A'\left(\frac{1}{2} \times 4, \frac{1}{2} \times 0\right)$ or $A'(2, 0)$

$B(2, 3) \rightarrow B'\left(\frac{1}{2} \times 2, \frac{1}{2} \times 3\right)$ or $B'\left(1, 1\frac{1}{2}\right)$

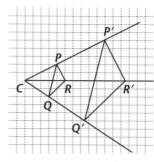

The transformation maps every point in segment AB to segment $A'B'$.

Example 2: What is the scale factor used to transform $\triangle PQR$ to the image $\triangle P'Q'R'$?

Every dilation has a point known as the center of dilation, which in this case is point C. $\triangle P'Q'R'$ is an enlargement of $\triangle PQR$, so the scale factor must be greater than 1. The distance from C to P' is $\frac{5}{2}$ times greater than the distance from C to P (the same is true for the distances from C to Q' and C to Q, and from C to R' and C to R), so we call this an enlargement with a scale factor of $\frac{5}{2}$.

A square with sides 3 centimeters long undergoes a dilation with a scale factor of 2.5.

1. How long is each side of the new image? _____

2. What is the area of the new image? _____

A triangle with a base of 3 centimeters and a height of 4 centimeters undergoes a dilation with a scale factor of 2.

3. What is the area of the original triangle? _____

4. What is the area of the new image? _____

5. If the scale factor of a dilation is n, what is the ratio of the area of the image to the area of the pre-image? _____

Basic Skills Practice

Scale and Dilations, Part 2

Read each question and circle the best answer.

1. Find the coordinates of the image points for $A(2, 2)$ and $B(6, 6)$ under a dilation with a scale factor 0.25. The origin is the center of dilation.

   A $A(0.5, 0.5), B(1.5, 1.5)$

   B $A(1, 0.5), B(1.5, 1.5))$

   C $A(-0.5, 0.5), B(1.5, -1.5)$

   D $A(2, 2), B(6, 4)$

2. The dotted line figure is a dilation of the solid line figure. Identify whether it is an enlargement or a reduction and find the scale factor.

   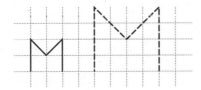

   F enlargement, scale factor of 3

   G reduction, scale factor of $\frac{1}{3}$

   H enlargement, scale factor of 2

   I reduction, scale factor of $\frac{1}{2}$

3. The dotted line figure is a dilation of the solid line figure. Identify whether it is an enlargement or a reduction and find the scale factor.

   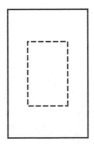

   A reduction, scale factor of $\frac{1}{2}$

   B enlargement, scale factor of 2

   C reduction, scale factor of $\frac{1}{4}$

   D enlargement, scale factor of 4

4. Fanny is drawing blueprints for a new house. On the blueprint, the entry hall measures 5 inches by 3.5 inches. If the scale used for the drawing is 0.5 in. = 1 ft, What are the dimensions of the actual entry hall?

   F 5 ft by 3.5 ft

   G 2.5 ft by 1.75 ft

   H 10 ft by 7 ft

   I 8 ft by 9 ft

5. Diana is going to enlarge a family picture. The original print is 4 inches by 6 inches. She is going to use a scale factor of 10. What are the dimensions of the enlargement?

   A 14 in. by 16 in.

   B 40 in. by 60 in.

   C 4 ft by 6 ft

   D 40 ft by 60 ft

6. Using a copy machine, a drawing of a tower that was 24 inches high was reduced to a tower that measured 8 inches high. What was the scale factor used for the reduction?

   F $\frac{1}{2}$

   G $\frac{1}{3}$

   H $\frac{1}{4}$

   I $\frac{1}{8}$

7. Given a triangle with vertices at $A(2, 2)$, $B(3, -1)$, and $C(1, 0)$, find the vertices of the image of $\triangle ABC$, using the origin as the center of dilation and a scale factor of 2.

   A $A(2, 4), B(4, 2), C(3, 6)$

   B $A(2, 1), B(-1, 3), C(0, 1)$

   C $A(4, 4), B(6, -2), C(2, 0)$

   D $A(4, 4), B(-2, 6), C(0, 2)$

Basic Skills Practice

Basic Skills Practice

Cumulative Review

Read each question and circle the best answer.

1. Pam mixes 2 quarts of window cleaner concentrate with 10 quarts of water. How many *gallons* of window cleaner does this make?

   A 3 gal

   B 4 gal

   C 12 gal

   D 20 gal

2. A student, 5 feet tall, casts a shadow 2 feet long. Find the height of a building that casts a shadow of 52 feet at the same time.

   F 21 ft

   G 52 ft

   H 104 ft

   I 130 ft

3. A stockholder received 4.575 shares as a dividend. Prior to the dividend, he owned 360.45 shares of the stock. How many shares of the stock does the stockholder now own?

   A 364.925 shares

   B 365.025 shares

   C 375.025 shares

   D 405.20 shares

4. In a scale drawing, a 280-ft tower is drawn 7 inches high. A building next to it is drawn 2 inches high. How tall is the actual building?

   F 140 ft

   G 120 ft

   H 100 ft

   I 80 ft

5. Find the circumference of the circle. (Use $\pi = 3.14$)

   8.75 m

   A 13.738 m

   B 27.475 m

   C 27.575 m

   D 54.95 m

6. Sally, age 43, has two sons named John and Rob. John is 7 years older than Rob. The sum of the sons' ages is less than Sally's age. Which inequality could be used to find J, the maximum age John could be?

   F $43 > J + 7 - J$

   G $2J - 7 < 43$

   H $43 - J > J + 7$

   I $J + 7 < 43$

7. If a pipe that is 12 feet 9 inches long is cut into 30 sections of equal length, how long is each section?

   A 0.67 in.

   B 4.3 in.

   C 5.1 in.

   D 6.7 in.

Basic Skills Practice
Cumulative Review

Read each question and circle the best answer.

1. Two tanks hold a total of 45 gallons of water. One tank holds 6 gallons more than twice the amount in the other. How much water is in the smaller tank?

   A 13 gal

   B 15 gal

   C 19 gal

   D 11 gal

2. What is the length of side *AB*?

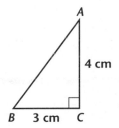

   F 4 cm

   G 3.5 cm

   H 5 cm

   I 7 cm

3. What is $\frac{2}{5}$ expressed as a decimal?

   A 0.04

   B 0.2

   C 0.25

   D 0.4

4. Find the mode for the following heights: 98 cm, 115 cm, 150 cm, 107 cm, 115 cm, 140 cm, 133 cm.

   F 98 cm

   G 115 cm

   H 123 cm

   I 858 cm

5. An international traveler is checking 7 pieces of luggage having a combined weight of 183.4 kilograms. What is the mean (average) weight of the pieces of luggage?

   A 26.2 kg

   B 26.7 kg

   C 27.31 kg

   D 34.7 kg

6. What is the surface area of a cube that has edges of 3 feet?

   F 9 ft^2

   G 18 ft^2

   H 27 ft^2

   I 54 ft^2

7. A pipe made of material that is 0.15 millimeters thick has an outside diameter of 3.575 millimeters. What is the inside diameter of the pipe?

   A 3.275 mm

   B 3.425 mm

   C 3.545 mm

   D 3.572 mm

8. Susan wants to rent a garden plot near her apartment. Which is the best estimate of the area of a rectangular plot that is 38 feet wide and 62 feet long?

   F 1500 ft^2

   G 2000 ft^2

   H 2400 ft^2

   I 2500 ft^2

Basic Skills Practice
Using Scientific Notation, Part 1

Scientific notation is a condensed way of expressing numbers. Very large or very small numbers often have many zeros that make them difficult to read and to use for calculations. A number written in scientific notation is written with two factors, a number between 1 and 10, but not including 10, and a power of 10.

Example 1: Write 3,270,000 in scientific notation.

$$3.27 \quad \rightarrow \quad \text{Write the first factor (a number between 1 and 10).}$$
$$10^6 \quad \rightarrow \quad \text{Write the power of 10.}$$
$$3,270,000 = 3.27 \times 10^6$$

The exponent is the number of places that the decimal point moved.
- If the original number is greater than 10, then the exponent is positive.
- If the original number is less than 1 then the exponent is negative.
- If the original number is between 1 and 10, then the exponent is 0.

Example 2: Write 0.0068 in scientific notation.

$$6.8 \quad \rightarrow \quad \text{Write the first factor.}$$
$$10^{-3} \quad \rightarrow \quad \text{Write the power of 10.}$$
$$0.0068 = 6.8 \times 10^{-3}$$

Write the missing exponent.

1. $345 = 3.45 \times 10^?$ _____

2. $5432 = 5.432 \times 10^?$ _____

3. $500,000 = 5 \times 10^?$ _____

4. $0.00038 = 3.8 \times 10^?$ _____

5. $0.21 = 2.1 \times 10^?$ _____

6. $42,000 = 4.2 \times 10^?$ _____

7. $854 = 8.54 \times 10^?$ _____

8. $0.0298 = 2.98 \times 10^?$ _____

9. $78 = 7.8 \times 10^?$ _____

10. $0.0000098 = 9.8 \times 10^?$ _____

Write in scientific notation.

11. 46 _____

12. 7900 _____

13. 800 _____

14. 0.000067 _____

15. 6,500,000 _____

16. 0.62 _____

17. 0.37 _____

18. 5276 _____

19. 4 _____

20. 856 _____

Basic Skills Practice

Using Scientific Notation, Part 2

Read each question and circle the best answer.

1. The distance of a galaxy from the Earth is expressed as 8.557×10^{10} miles. Which is another way to express this amount?

   A 75,530,000 mi

   B 855,7000,000 mi

   C 8,557,000,000 mi

   D 85,570,000,000 mi

2. The diameter of a particle is 1.2×10^{-6} millimeters. Express in standard notation.

   F 0.00000012 mm

   G 0.0000012 mm

   H 0.00012 mm

   I 0.0012 mm

3. The number of grains of sand on a section of a beach is 6.85×10^9. Express in standard notation.

   A 6,850,000,000

   B 685,000,000

   C 68,500,000

   D 685,000

4. The thickness of a soap bubble is about 0.00002 meters. What is this in scientific notation?

   F 0.2×10^{-4} m

   G 2×10^{-5} m

   H 0.2×10^{-6} m

   I 20×10^{-6} m

5. A convention attracted 1.17×10^4 left-handed people. Express in standard notation.

   A 117

   B 1170

   C 11,700

   D 1,170,000

6. A collector owns 11,231 baseball cards. Express in scientific notation.

   F 1.1231×10^{-4}

   G 11.231×10^{-3}

   H 1.1231×10^2

   I 1.1231×10^4

7. A chemical compound contains 0.0428 milligrams of phosphate. Express in scientific notation.

   A 4.28×10^{-3} mg

   B 4.28×10^{-2} mg

   C 4.28×10^{-1} mg

   D 4.28×10^3 mg

8. The distance from the Earth to the moon is about 239,000 miles. How is this measurement expressed in scientific notation?

   F 2.39×10^6 mi

   G 2.39×10^5 mi

   H 23.9×10^5 mi

   I 2×10^6 mi

Solve each problem.

9. There were 4.544×10^4 fans at a concert. Express in standard notation. _____

10. A jet flew at an altitude of 1.4×10^4 feet. Express in standard notation. _____

Basic Skills Practice

Effects of Dimensional Changes, Part 1

Charlie works for a company that builds cardboard boxes. Their most popular box is a cube whose sides measure 2 feet, whose surface area is 24 square feet, and whose volume is 8 cubic feet. Charlie wants to design a box with dimensions twice as big.

Example 1: Find the surface area of the new box.
Use the formula for the surface area of a cube, $S = 6s^2$,
to find the surface area of the new box.
$S = 6(2 \times 2)^2$ Multiply the dimension of the sides by 2.
$S = 6(4)^2$
$S = 6(16)$
$S = 96$ ft^2

Example 2: Find the volume of the new box.
Use the formula for the volume of a cube, $V = s^3$,
to find the volume of the new box.
$V = (2 \times 2)^3$ Multiply the dimension of the sides by 2.
$V = (4)^3$
$V = 64$ ft^3

Use the same method to find the dimensional changes for rectangular prisms and cylinders.

Solve each problem.

1. A rectangular prism has sides of 3 inches, 4 inches, and 5 inches. What are the new dimensions of the prism if the sides are reduced by one-half?   _____

2. A cylinder has a radius of 1 meter and a height of 2.5 meters. What is the volume of a cylinder with dimensions three times larger?   _____

3. A cube has sides that measure 6 inches. What is the surface area of the cube if its dimensions are reduced by one-third?   _____

4. A cylinder has a radius of 12 inches and a height of 4 inches. What is the volume of the cylinder if its dimensions are reduced by one-fourth?   _____

5. A rectangular prism has sides of 3 inches, 4 inches, and 5 inches. What is the volume of the prism if its dimensions are doubled?   _____

Travis makes planters in the shape of cubes. He usually makes planters with sides that measure 6 inches, but he now wants to make a planter with sides that measure 4 inches.

6. What is the volume of the planter Travis usually makes?   _____

7. What is the volume of the planter Travis wants to make?   _____

Basic Skills Practice

Effects of Dimensional Changes, Part 2

Read each question and circle the best answer.

1. A cube has a surface area of 24 square inches. What is the area of the cube if each side is doubled?

   A 54 in.^{2}

   B 96 in.^{2}

   C 30 in.^{2}

   D 12 in.^{2}

2. How much more volume does a cube with sides measuring 5 inches have than a cube with sides measuring 3 inches?

   F 20 in.^{3}

   G 98 in.^{3}

   H 150 in.^{3}

   I 61 in.^{3}

3. Joy is in charge of buying containers for the school's recycling center. Of the two cylindrical containers that she can choose, one container has dimensions that are twice as big as the other container. If the smaller cylinder has a height of 2 meters and a radius of 1 meter, what is the volume of the larger cylinder?

   A 12π m^3

   B 16π m^3

   C 24π m^3

   D 40π m^3

4. How much more volume does the larger cylinder have than the smaller cylinder?

   F 16π m^3

   G 2π m^3

   H 14π m^3

   I 8π m^3

5. A rectangular prism has dimensions of 2 inches, 4 inches, and 6 inches. What is the surface area of the prism if the dimensions are multiplied by 1.5?

   A 118 in.^{2}

   B 108 in.^{2}

   C 198 in.^{2}

   D 218 in.^{2}

6. What is the volume of the larger prism?

   F 162 in.^{3}

   G 118 in.^{3}

   H 198 in.^{3}

   I 218 in.^{3}

7. How much more volume does the larger prism hold than the smaller prism?

   A 44 in.^{3}

   B 114 in.^{3}

   C 88 in.^{3}

   D 108 in.^{3}

Solve each problem.

8. The ratio of the sides of two rectangular prisms is 1:3. What is the ratio of their surface areas? _____

9. What is the ratio of their volumes? _____

Basic Skills Practice

Cumulative Review

Read each question and circle the best answer.

1. A cube has sides that measure 6 inches. By how much would its volume increase if the dimensions were multiplied by 1.5?

A 513 in.^{3}

B 216 in.^{3}

C 529 in.^{3}

D 100 in.^{3}

2. By how much would the surface area increase for the cube in Exercise 1?

F 118 in.^{2}

G 270 in.^{2}

H 486 in.^{2}

I 529 in.^{2}

3. The dimensions of a rectangular prism are 6 inches by 12 inches by 15 inches. What would the dimensions be if they were reduced by $\frac{1}{3}$?

A 3 in. $\times$ 6 in. $\times$ 9 in.

B 2 in. $\times$ 4 in. $\times$ 5 in.

C 4 in. $\times$ 8 in. $\times$ 10 in.

D 3 in. $\times$ 9 in. $\times$ 12 in.

4. By how much would the volume decrease for the rectangular prism in Exercise 3?

F 240 in.^{3}

G 360 in.^{3}

H 540 in.^{3}

I 760 in.^{3}

5. Find the median for the following scores: 9, 4, 8, 9, 10, 11, 4, 0, and 6.

A 6.7

B 61

C 9

D 8

6. The radius of a cylinder is 4 inches. The height is 24 inches. What would the volume be if its dimensions were reduced by three-fourths?

F 6π in.^{3}

G 12π in.^{3}

H 20π in.^{3}

I 24π in.^{3}

7. What would be the resulting cylinder's lateral surface area in Exercise 6?

A 6π in.^{2}

B 9π in.^{2}

C 12π in.^{2}

D 24π in.^{2}

8. Given a square with vertices at $A(2, 1)$, $B(2, 4)$, $C(5, 4)$, and $D(5, 1)$, find the vertices of the image of $ABCD$ by using the origin as the center of dilation and a scale factor of 2.

F $A(2, 1)$, $B(2, 4)$, $C(5, 4)$, $D(5, 1)$

G $A(4, 2)$, $B(4, 8)$, $C(10, 8)$, $D(10, 2)$

H $A(4, 1)$, $B(4, 4)$, $C(10, 4)$, $D(10, 1)$

I $A(2, 2)$, $B(2, 8)$, $C(5, 8)$, $D(5, 2)$

Solve each problem.

9. The dimensions of a rectangular prism are 15 inches by 20 inches by 25 inches.

What is the volume, in cubic inches, if the dimensions were multiplied by $\frac{1}{5}$? _____

Basic Skills Practice

Counting Outcomes and Tree Diagrams, Part 1

A sample space is a listing of all of the possible outcomes in a given situation. You can use a tree diagram to show all of the possible outcomes. Each "branch" represents a choice or possible outcome.

Example 1: A private high school offers its students a variety of uniform choices. Students can wear khaki or navy blue pants. The shirt can be white or light blue, and the tie can be black or blue. Make a diagram to show all of the possible uniforms.

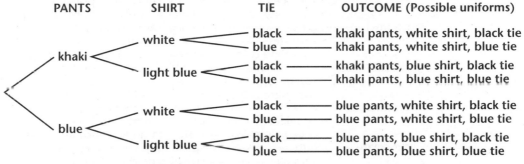

PANTS   SHIRT   TIE   OUTCOME (Possible uniforms)

khaki	white	black	khaki pants, white shirt, black tie
		blue	khaki pants, white shirt, blue tie
	light blue	black	khaki pants, blue shirt, black tie
		blue	khaki pants, blue shirt, blue tie
blue	white	black	blue pants, white shirt, black tie
		blue	blue pants, white shirt, blue tie
	light blue	black	blue pants, blue shirt, black tie
		blue	blue pants, blue shirt, blue tie

There are 8 possible uniforms.

To find the total number of possible outcomes of a series of events, multiply the number of ways each event can occur.

Example 2: A store issues identification codes to its employees. The codes start with a letter and are followed by a two-digit number. How many different codes are possible?

$$\text{letter (A to Z)} \quad \text{digit (0 to 9)} \quad \text{digit (0 to 9)}$$
$$26 \quad \times \quad 10 \quad \times \quad 10$$

2600 different codes are possible.

1. The lunch special at Stuart's Cafe includes soup, salad, and dessert. Complete the tree diagram to show all of the possible lunch special combinations.

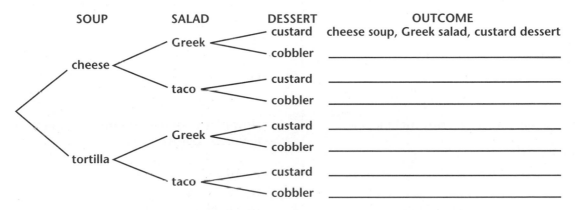

SOUP   SALAD   DESSERT   OUTCOME

cheese	Greek	custard	cheese soup, Greek salad, custard dessert
		cobbler	_____
	taco	custard	_____
		cobbler	_____
tortilla	Greek	custard	_____
		cobbler	_____
	taco	custard	_____
		cobbler	_____

2. How many possible lunch specials are there? _____

NAME _____ CLASS _____ DATE _____

Basic Skills Practice

Counting Outcomes and Tree Diagrams, Part 2

Read each question and circle the best answer.

The frozen yogurt shop is having a back-to-school special. You can choose from three flavors of yogurt, two kinds of hot sauces, and three kinds of dry toppings to make a sundae for the special price of $1.75. Refer to the tree diagram to answer Exercises 1–6.

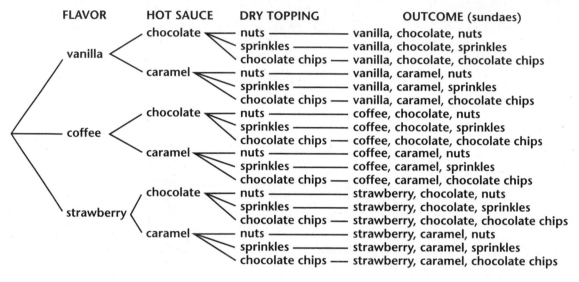

1. How many sundaes include coffee-flavored yogurt?

   A 2
   B 4
   C 6
   D 8

2. How many sundaes have sprinkles as a dry topping?

   F 7
   G 6
   H 5
   I 4

3. How many sundaes do **not** include nuts?

   A 9
   B 10
   C 12
   D 14

4. How many sundaes include vanilla-flavored yogurt with caramel sauce?

   F 1
   G 3
   H 6
   I 9

5. How many sundaes that include chocolate sauce can be made?

   A 9
   B 7
   C 6
   D 3

6. How many sundaes do **not** include vanilla-flavored yogurt?

   F 10
   G 12
   H 15
   I 18

Basic Skills Practice

Basic Skills Practice

Probability, Part 1

The probability of an event is the likelihood that the event will occur. When an event is certain to occur, it has a probability of 1. When an event is certain not to occur, it has a probability of 0.

When an event is equally likely to occur or not occur, it has a probability of $\frac{1}{2}$, or 0.5.

The probability (P) that an outcome will occur is the ratio of the number of successful outcomes to the number of possible outcomes:

$$P = \frac{\text{successful outcomes}}{\text{possible outcomes}}$$

When you toss a coin, there are two possible outcomes, heads or tails.

The chances of tossing tails are $\frac{\text{successful outcomes}}{\text{possible outcomes}} = \frac{\text{tails}}{\text{heads or tails}} = \frac{1}{2}$.

Example 1: When rolling a number cube, what is the probability of getting a 3?
successful outcomes: 3
possible outcomes: 1, 2, 3, 4, 5, 6
$$P = \frac{\text{number of successful outcomes}}{\text{number of possible outcomes}} = \frac{1}{6}$$

Example 2: When rolling a number cube, what is the probability of getting an even number?
successful outcomes: 2, 4, 6
possible outcomes: 1, 2, 3, 4, 5, 6
$$P = \frac{3}{6} = \frac{1}{2}$$

Exercises 1–4 refer to one roll of a number cube.

1. What is the probability of getting an even number?   _____

2. What is the probability of getting a 3 or a 5?   _____

3. What is the probability of getting a multiple of 3?   _____

4. What is the probability of getting 7?   _____

Exercises 5–8 refer to a bag filled with 3 yellow balls, 4 green balls, 2 blue balls, and 1 orange ball.

5. What is the probability of picking a red ball out of the bag?   _____

6. What is the probability of picking a blue or yellow ball?   _____

7. What is the probability of picking an orange ball?   _____

8. What is the probability of picking a green ball?   _____

9. What is the probability of picking a green or orange ball?   _____

Basic Skills Practice

Probability, Part 2

Read each question and circle the best answer.

Use the spinner below to answer Exercises 1–5.

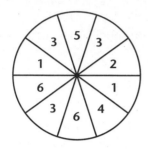

1. Find the probability of spinning an odd number.

   A $\frac{1}{5}$

   B $\frac{2}{5}$

   C $\frac{3}{5}$

   D $\frac{4}{5}$

2. Find the probability of spinning a 5 or a 6.

   F $\frac{1}{10}$

   G $\frac{2}{10}$

   H $\frac{3}{10}$

   I $\frac{7}{10}$

3. Find the probability of spinning a 1.

   A $\frac{4}{5}$

   B $\frac{3}{5}$

   C $\frac{2}{5}$

   D $\frac{1}{5}$

4. Find the probability of spinning a 3 or a 4.

   F $\frac{1}{10}$

   G $\frac{4}{10}$

   H $\frac{2}{10}$

   I $\frac{5}{10}$

5. Find the probability of spinning an even number.

   A $\frac{1}{10}$

   B $\frac{3}{10}$

   C $\frac{4}{10}$

   D $\frac{7}{10}$

6. A letter from the word GUARANTEE is picked at random. What are the odds of picking a vowel?

   F $\frac{7}{9}$

   G $\frac{5}{9}$

   H $\frac{4}{9}$

   I $\frac{2}{9}$

7. What is the probability of getting an odd number in one roll of a number cube?

   A $\frac{1}{10}$

   B $\frac{3}{10}$

   C $\frac{1}{2}$

   D $\frac{7}{10}$

Basic Skills Practice
Cumulative Review

Read each question and circle the best answer.

1. The Corner Diner offers a "design your own lunch" special with soup, salad, and dessert. If there are 2 soups, 3 salads, and 4 desserts, how many different lunch specials can be designed?

   A 9

   B 10

   C 20

   D 24

2. Mr. Baehr said he would give a surprise test one day next week. What is the probability of him giving the test on Monday?

   F $\frac{1}{2}$

   G $\frac{1}{3}$

   H $\frac{1}{4}$

   I $\frac{1}{5}$

3. What is 0.0037 expressed in scientific notation?

   A 3.7×10^{-2}

   B 3.7×10^{-3}

   C 37×10^{-4}

   D 37×10^{-5}

4. Given $\triangle FGH$ with vertices at $F(-2, 2)$, $G(4, 2)$, and $H(0, -2)$, what are the vertices of its image under a dilation with its center at the origin and a scale factor of $\frac{3}{2}$?

   F $F(-3, 3)$, $G(6, 3)$, $H(0, -3)$

   G $F(3, 3)$, $G(6, 2)$, $H(0, -3)$

   H $F(-3, 3)$, $G(6, 3)$, $H(1, -3)$

   I $F(-5, 3)$, $G(6, 2)$, $H(0, -3)$

5. What relationship best describes $\triangle ABC$ and its translation, $\triangle A'B'C'$?

   A They are similar triangles.

   B They are congruent triangles.

   C They are parallel triangles.

   D They are rotated triangles.

6. In one roll of a number cube, what is the probability of rolling a 5 or a 3?

   F $\frac{1}{2}$

   G $\frac{1}{3}$

   H $\frac{1}{4}$

   I $\frac{1}{6}$

Solve each problem.

7. Write the missing exponent in $0.0000087 = 8.7 \times 10^{\blacksquare}$. _____

8. Given a bag with the names of 10 male students and 12 female students, what is the probability of reaching into the bag and picking a female student's name? _____

9. What is the probability of getting an even number on one roll of a six-sided number cube? _____

Basic Skills Practice

Cumulative Review

Read each question and circle the best answer.

1. Write a mathematical sentence for "a number tripled and decreased by 17."

   A $3x - 17$

   B $17 - 3x$

   C $3x + 17$

   D $17 + 3x$

2. The soccer coach is 5 years older than twice the age of the youngest team member. Write the coach's age as an algebraic expression.

   F $5 - 2x$

   G $2x + 5$

   H $5x$

   I $5x + 2$

3. The area of a square field is 400 square feet. What is the length of each side?

   A 100 ft

   B 40 ft

   C 20 ft

   D 10 ft

4. Which number sentence matches the number line?

   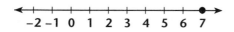

   F $x \leq 7$

   G $x \geq -2$

   H $x = 7$

   I $x < -2$

5. Find the volume of a cylinder with a radius of 5 feet and a height of 8 feet. Use 3.14 for π.

   A 628 ft³

   B 528 ft³

   C 125.6 ft³

   D 31.4 ft³

6. A storage bin has a volume of 693 cubic feet. It is 9 feet long and 11 feet wide. How high is the bin?

   F 5 ft

   G 7 ft

   H 9 ft

   I 11 ft

7. How many outfits can be made with 2 color choices for pants, 3 color choices for shirts, and 3 color choices for vests?

   A 8

   B 9

   C 12

   D 18

8. The Corner Paint Shop sells 4 brands of house paint. Each brand has 20 colors to choose from. How many different choices of paint does the paint shop sell?

   F 16

   G 24

   H 40

   I 80

Solve each problem.

9. A rectangular barn has a width of 10 feet, a length of 12 feet, and a height of 9 feet. Find the volume of the barn. _____

10. What is 15.39 rounded to the nearest tenth? _____

Basic Skills Practice
Cumulative Review

Read each question and circle the best answer.

1. Find the equation that matches the number line.

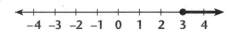

   A $x + 5 = 5$

   B $x - 2 = -3$

   C $2x + 2 = 8$

   D $16 - 8 = 3x$

2. Find the inequality that matches the number line.

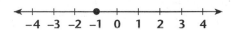

   F $6n + 4 \leq 10$

   G $3n \geq 9$

   H $n \leq -6$

   I $n > 0$

3. What is the volume of a cylinder with a radius of 5 inches and a height of 8 inches?

   A 150π in.^{3}

   B 200π in.^{3}

   C 500π in.^{3}

   D 1500π in.^{3}

4. Find the graph that matches the inequality.
   $$2x + 3 < 5$$

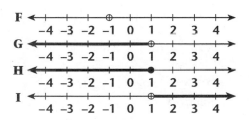

5. Triangle *ABC* is similar to triangle *DEF*. What is the value of *x*?

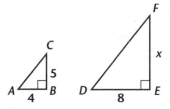

   A 12

   B 10

   C 9

   D 6

6. Joy's Ice Cream Parlor has 13 flavors of ice cream, 3 hot sauces, and 4 dry toppings. Using 1 flavor of ice cream, 1 hot sauce, and 1 dry topping, how many different sundaes can you make?

   F 20

   G 25

   H 120

   I 156

7. Susan is pregnant with twins. What is the probability that she will have one boy and one girl?

   A 0

   B $\frac{1}{3}$

   C $\frac{1}{2}$

   D $\frac{2}{3}$

Basic Skills Practice
Cumulative Review

Read each question and circle the best answer.

1. What happens to the area of a circle when the radius is doubled in size?

   A The area remains the same.

   B The area doubles.

   C The area triples.

   D The area quadruples.

2. What happens to the volume of a cylinder when the height is halved?

   F The volume remains the same.

   G The volume is one-half of what it was before.

   H The volume is double what it was before.

   I The volume is one-third of what it was before.

3. To solve the equation $5a + 2 = 17$, you need to—

   A add 2 to both sides and then divide both sides by 3

   B subtract 2 from both sides and then divide both sides by 5

   C divide both sides by 5

   D divide both sides by 17

4. According to the scale on a map, 1 inch represents 5 miles. The distance between the museum and the airport is 3.5 inches on the map. What is the actual distance in miles?

   F 15 mi

   G 17.5 mi

   H 19.5 mi

   I 21 mi

5. The ratio of the size of an object in a drawing to the size of the actual object is called a(n)—

   A scale

   B reduction

   C enlargement

   D rotation

6. Amalia is carpeting two bedrooms with the same type of carpet. One room is 8 feet by 10 feet, and the other is 9 feet by 12.5 feet. How many square feet of carpeting does she need?

   F 190.5 ft^2

   G 192.5 ft^2

   H 205 ft^2

   I 215 ft^2

7. Given similar triangles *ABC* and *DEF*, what is true about $\angle A$ and $\angle D$?

   A They are different.

   B They are complementary.

   C They are congruent.

   D They are similar.

8. The figures below represent—

   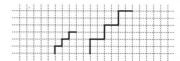

   F a rotation

   G a translation

   H a reflection

   I a dilation

Solve.

9. What is the probability of getting a 7 on one roll of a six-sided number cube? _____

Basic Skills Practice

Interpreting Bar and Circle Graphs, Part 1

Graphs provide a visual representation of data. A bar graph is a useful tool to compare data. A double bar graph can be used to compare two sets of data. Bar graphs should have a title and a scale.

The horizontal bar graph at right shows the cost of long distance calls to 4 different countries.

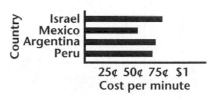

Example 1: Which country is the most expensive to call?
To find the most expensive country, look for the longest bar which is Israel.

The double bar graph at right shows the favorite types of food for 2 different age groups.

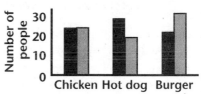

Example 2: Which food is equally popular with both groups?
To find the food that is equally popular, look for bars that have the same height which is chicken.

A circle graph is useful to show data as part of a whole. A circle graph should have a title, and each portion of the circle should be labeled and quantified as a part of a whole (as a fraction or as a percent).

The circle graph at right shows the distribution of Parker's expenses.

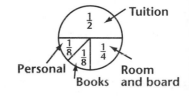

Example 3: What fraction of Parker's expenses go toward books?
To find the answer, look for the portion of the circle that is labeled *Books,* and read the fraction that is written for it, $\frac{1}{8}$.

The school cafeteria conducted a survey to see how students felt about 4 different kinds of food. The results are shown in the bar graph on the right.

Use the bar graph to answer Exercises 1–5.

1. How many students like hamburger? _____

2. How many students like chicken? _____

3. How many more students like turkey than fish? _____

4. How many students answered the survey? _____

5. Which food item was the favorite among the students surveyed? _____

Basic Skills Practice

Interpreting Bar and Circle Graphs, Part 2

Read each question and circle the best answer.

Use the bar graph to answer Exercises 1–4.

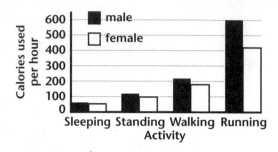

1. How many calories does a female use during 1 hour of standing?

   A 200

   B 100

   C 150

   D 50

2. How many calories does a male use by walking for 1 hour?

   F 180

   G 220

   H 440

   I 600

3. How many more calories does a male use than a female during 1 hour of running?

   A 100

   B 150

   C 200

   D 300

4. Which activity uses the fewest calories in 1 hour?

   F walking

   G standing

   H sleeping

   I running

Use the bar graph to answer Exercises 5–8.

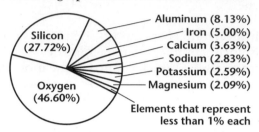

5. Which element makes up almost half of the Earth's crust?

   A iron

   B oxygen

   C silicon

   D aluminum

6. What percent of the Earth's crust is made up of sodium?

   F 2.83%

   G 3.63%

   H 27.72%

   I 30.52%

7. Oxygen and silicon make up 74.32% of the Earth's crust. What percent do the rest of the elements represent?

   A 35.68%

   B 25.68%

   C 15%

   D 2%

8. What percent of the Earth's crust is made up of elements that are present in less than 1%?

   F 1%

   G 1.41%

   H 3%

   I 4.3%

Basic Skills Practice

Interpreting Line Graphs and Scatter Plots, Part 1

Line graphs are made up of connected points. Each point connects a horizontal value and a vertical value.

Example 1: The graph shows the number of cars sold by month over a 4-month period. By how much did sales increase or decrease between March and April?

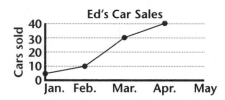

30 cars were sold in March. 40 cars were sold in April. Sales increased by 10 cars between March and April.

Scatter plots also contain points that connect a horizontal value with a vertical value. Scatter plots are used to find trends between two sets of data that show positive, negative, or no correlation.

Positive correlation
Both sets of data tend to increase together.

Negative correlation
One set of data tends to increase as the other decreases.

No correlation
The sets of data are not related.

Solve each problem.

The graph at right shows the enrollment at a college.

1. How many students were enrolled in 1980? _____

2. In what year were 8000 students enrolled? _____

3. How many students were enrolled in 1985? _____

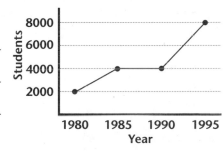

The scatter plot at right shows the relationship between the temperature and the distance that several athletes were able to run in a set amount of time.

4. As temperature increases, the distance tends to _____.

5. What correlation does the scatter plot show? _____

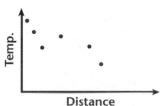

Basic Skills Practice

Interpreting Line Graphs and Scatter Plots, Part 2

Read each question and circle the best answer.

The following graph shows the temperatures in degrees Fahrenheit between 7 A.M. and 11 A.M. at a ballpark.

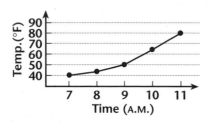

1. What is the highest temperature recorded?

   A 10°F

   B 11°F

   C 80°F

   D 90°F

2. When was the lowest temperature recorded?

   F 7 A.M.

   G 8 A.M.

   H 9 A.M.

   I 10 A.M.

3. What was the temperature at 9 A.M.?

   A 40°F

   B 50°F

   C 60°F

   D 70°F

Greg took 4 tests. His scores are recorded on the line graph below.

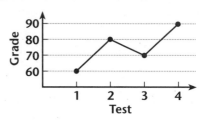

4. On which test did Greg receive the highest score?

   F 1

   G 2

   H 3

   I 4

5. On which test did Greg receive a score of 80?

   A 1

   B 2

   C 3

   D 4

6. What score did Greg receive on test 3?

   F 60

   G 70

   H 80

   I 90

Solve each problem.

7. What is the approximate height for a person weighing 70 pounds? _____

8. What is the approximate weight for a person who has a height of 55 inches? _____

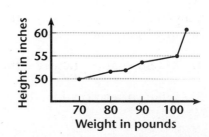

Basic Skills Practice

Cumulative Review

Read each question and circle the best answer.

The following line graph charts the change in the price of a certain sculpture over time.

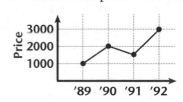

1. What was the cost of the sculpture in 1989?

   A $1000

   B $1500

   C $2000

   D $3000

2. In what year did the sculpture cost $3000?

   F 1989

   G 1990

   H 1991

   I 1992

3. By how much did the sculpture's price increase between 1989 and 1992?

   A $3000

   B $1000

   C $1500

   D $2000

4. In what year did the sculpture cost the least?

   F 1989

   G 1990

   H 1991

   I 1992

5. What was the change in price of the sculpture between 1990 and 1991?

   A An increase of $500

   B A decrease of $500

   C An increase of $1000

   D A decrease of $1000

Use the graph to answer Exercises 6–8.

The line graph charts the change in average attendance at a certain hockey team's games.

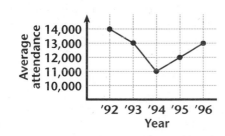

6. Which year had the lowest average attendance at the team's games? _____

7. Which year had the highest average attendance at the team's games? _____

8. What was the average attendance in 1994? _____

9. By how much did average attendance decline over the period 1992–1996? _____

Basic Skills Practice

Interpreting Venn Diagrams and Tables, Part 1

Venn diagrams represent relationships among events. In a Venn diagram, the rectangle represents a sample group and the circles represent events within that group.

Example 1: A survey was conducted to see if people watched either of two movies.

S = set of 100 people surveyed
A = set of people who watched movie A
B = set of people who watched movie B

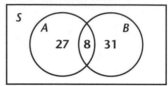

How many people watched only movie A?
The number in circle A that does not overlap circle B is 27. Therefore 27 people watched only movie A.

How many people watched movie A?
The overlapping region includes people who watched both movie A and movie B. Therefore 8 plus 27 or 35 people in total watched movie A.

Tables organize information by grouping data.

Example 2: Use the table below to find who had the best golf score on Saturday. Remember that in golf the lowest score is the best.

Golf Scores

	Tim	Brad
Friday	72	68
Saturday	78	81

Since Tim scored a 78 and Brad scored 81, Tim had the best score on Saturday.

Use the graph to answer the following exercises:

A newspaper surveyed 500 people to determine how many people own cars and how many people own trucks. The results are shown in the Venn diagram.

L = set of 500 people surveyed
C = set of car owners
T = set of truck owners

1. How many people own only cars? _____

2. How many people own only trucks? _____

3. How many people own ***neither*** cars ***nor*** trucks? _____

4. How many people own both cars and trucks? _____

The table at right shows the elevation of several mountains.

5. What is the elevation of Mt. Hardy in feet? _____

6. What is the elevation of Mt. Price in meters? _____

Elevation

	Feet	Meters
Mt. Hardy	17,330	5282
Mt. Price	14,781	4505
Mt. Howard	15,526	4732

Basic Skills Practice

Interpreting Venn Diagrams and Tables, Part 2

Read each question and circle the best answer.

A questionnaire asked 200 janitors which cleaning product they used.

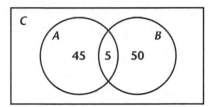

C = the set of 200 janitors surveyed
A = the set of janitors who used product A
B = the set of janitors who used product B

1. How many janitors used only product A?

    A 45
    B 50
    C 55
    D 60

2. How many janitors used only product B?

    F 45
    G 50
    H 55
    I 60

3. How many janitors used both products?

    A 5
    B 10
    C 55
    D 100

4. How many used **neither** product?

    F 5
    G 10
    H 55
    I 100

The table shows the monthly income and expenses for 4 students.

	Income	Expenses
Leon	$150	$75
Ben	$200	$210
Jay	$250	$240
Kara	$50	$10

5. Which student has the highest income?

    A Leon
    B Ben
    C Jay
    D Kara

6. Which student has the least expenses?

    F Leon
    G Ben
    H Jay
    I Kara

7. After expenses, how much money does Jay have left each month?

    A $10
    B $40
    C $75
    D $490

8. Which student has greater expenses than income?

    F Leon
    G Ben
    H Jay
    I Kara

Basic Skills Practice
Cumulative Review

Read each question and circle best answer.

1. A large bag of dog food weighs almost 15 pounds. To the nearest hundred pounds, how much will 225 bags weigh?

   A 3300 lb
   B 3400 lb
   C 3500 lb
   D 3600 lb

2. A fabric shop sells ribbon in 4 different widths. They are arranged on shelves in order of width in inches. Which group of fractions shows the correct order from greatest to least width?

   F $\frac{1}{2}, \frac{2}{9}, \frac{11}{18}, \frac{5}{12}$

   G $\frac{2}{9}, \frac{5}{12}, \frac{1}{2}, \frac{11}{18}$

   H $\frac{11}{18}, \frac{1}{2}, \frac{5}{12}, \frac{2}{9}$

   I $\frac{5}{12}, \frac{11}{18}, \frac{2}{9}, \frac{1}{2}$

3. The wavelength of red light is approximately 0.00000068 meters. Find the correct expression of this number in scientific notation.

   A 68×10^8 m
   B 68×10^{-8} m
   C 6.8×10^7 m
   D 6.8×10^{-7} m

4. Find the equation that is equivalent to the following statement: The difference of three times x and two times y is 42.

   F $3 + x - 2 + y = 42$
   G $3x + 2y = 42$
   H $3x - 2y = 42$
   I $3x - 2y + 42 = 0$

5. A bag contains 14 marbles. There are 8 white marbles, 4 red marbles, and 2 blue marbles. One marble is drawn from the bag. Find the probability that it is a red marble.

   A 1 out of 7
   B 2 out of 7
   C 4 out of 7
   D 6 out of 7

6. To make a lion costume, $3\frac{1}{4}$ yards of light brown felt was purchased and $\frac{7}{8}$ yards of dark brown felt was purchased. What was the combined yardage of the 2 colors of felt?

   F $3\frac{1}{8}$ yd

   G $3\frac{2}{3}$ yd

   H $3\frac{3}{4}$ yd

   I $4\frac{1}{8}$ yd

Solve each problem.

7. A rectangular swimming pool is 25 meters long and 18 meters wide. The average depth of the pool is 3 meters. What is the volume of the pool in meters? _____

8. What is the surface area, in square meters, of the pool referred to in Exercise 7? Include a cover to go over the top of the swimming pool. _____

Basic Skills Practice

Basic Skills Practice

Cumulative Review

Read each question and circle the best answer.

1. Jack needs 24 feet of fabric to make a banner for a party. How many **yards** of fabric does he need?

  A 8 yd

  B 12 yd

  C 48 yd

  D 72 yd

2. The approximate braking distance in feet necessary to stop a car is given by the formula

$$D = 0.05R^2,$$

where R is the speed of the car. What is the approximate braking distance for a car traveling 60 miles per hour?

  F 30 ft

  G 90 ft

  H 180 ft

  I 900 ft

3. Jacob bought 5 acres of land outside the city. He plans on donating $\frac{2}{3}$ of this land for a park. How many acres of land does he plan on donating?

  A $2\frac{1}{2}$ acres

  B $3\frac{1}{4}$ acres

  C $3\frac{1}{3}$ acres

  D $4\frac{1}{4}$ acres

4. Find the equation equivalent to this equation.

$$-5 + x = 8$$

  F $x = 8 + 5$

  G $x = 8 - 5$

  H $5 = -x + 8$

  I $-5 = -x - 8$

5. Kirby bought a purse at a discount of 25%. She saved $14.00. What was the original price of the purse?

  A $17.00

  B $35.00

  C $56.00

  D $560.00

6. The ratio of girls to boys on Savannah's soccer team is 2 to 3. If there are 12 boys on the team, how many girls are there?

  F 6

  G 8

  H 10

  I 18

7. Given a right triangle with sides a and b and a hypotenuse of 30 centimeters, which of the following equations is correct?

  A $a^2 + b^2 = 30$

  B $a^2 - 30 = b^2$

  C $a^2 + b^2 = 30^2$

  D $30 + b^2 = a^2$

8. Each of 5 friends intends to pay an equal share of the expenses for a camping trip. The cost per day for the campsite is $18.80. There is also an additional cost of $5.80 per vehicle. The 5 friends drove in 2 vehicles. How much should each friend pay for 1 day of camping, including the vehicle charge?

  F $6.08

  G $5.68

  H $5.98

  I $6

Basic Skills Practice

Cumulative Review

Read each question and circle the best answer.

1. Michael wants to find the mean (average) and range of his algebra test scores for the semester. The scores were 87, 92, 77, 90, 95, and 75. Find the mean and range.

   A mean of 85, range of 20

   B mean of 86, range of 20

   C mean of 87, range of 17

   D mean of 88, range of 17

2. Kara earned $40.00 last week. She spent $12.00 on a T-shirt. What percent of her earnings did she spend on the T-shirt?

   F 30%

   G 40%

   H 33%

   I 25%

3. Find the fraction in simplest terms that is equivalent to 40%.

   A $\dfrac{40}{100}$

   B $\dfrac{20}{50}$

   C $\dfrac{10}{25}$

   D $\dfrac{2}{5}$

4. Shawna takes photographs and then enlarges them. A photograph that is 5.6 inches wide and 8 inches long will be enlarged to a photograph that is 20 inches long. Find the proportion that can be used to find W, the width of the enlarged photograph.

   F $\dfrac{W}{8} = \dfrac{5.6}{20}$

   G $\dfrac{20}{5.6} = \dfrac{W}{8}$

   H $\dfrac{20}{W} = \dfrac{5.6}{8}$

   I $\dfrac{5.6}{8} = \dfrac{W}{20}$

5. Which expression is equivalent to this expression?
   $$3 + (4x)^2$$

   A $3 + 4x^2$

   B $12x^2$

   C $3 + 12x^2$

   D $3 + 16x^2$

6. Solve $10x - 80 = -100$.

   F $x = 2$

   G $x = -2$

   H $x = 18$

   I $x = -18$

7. Tom wants to carpet a bedroom with dimensions of 9 feet by 11 feet and a hallway with dimensions of 4 feet by 12 feet. What is the total amount of carpet he will need?

   A 36 ft²

   B 72 ft²

   C 147 ft²

   D 294 ft²

8. Members of a high school band have a red jacket, a blue jacket, a white shirt, a red shirt, and a blue shirt.

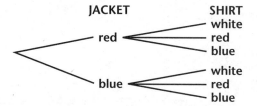

   Use the diagram to determine the number of different combinations of uniforms that the band members can wear.

   F 2

   G 4

   H 6

   I 8

Basic Skills Practice

Cumulative Review

Read each question and circle the best answer.

1. Find the missing numbers in the pattern.
   45, 40, 37, 32, _____, _____, 21, 16

   A 17, 14
   B 28, 25
   C 29, 26
   D 29, 24

2. Susie made a gift list for her family. The list includes the following amounts (including tax): $45.20, $33.15, $40.08, and $21.98. If she has budgeted $150.00 for gifts, how much money does she have left?

   F $8.59
   G $9.59
   H $10.59
   I $11.59

3. The value of Debra's pottery when she bought it 10 years ago was $55. This year, its value is $75. What is the approximate percent increase in its value?

   A 26%
   B 36%
   C 42%
   D 52%

4. Given $l \parallel m$ and $\angle A = 115°$, what is the measure of $\angle B$?

   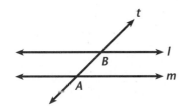

   F 35°
   G 65°
   H 115°
   I 125°

5. What is the sum of $0.6509 + 2.548$ rounded to the nearest tenth?

   A 3.199
   B 3.19
   C 3.2
   D 3.0

6. The perimeter of a square swimming pool is 702.25 meters. What is the approximate length of its sides?

   F 17.56 m
   G 175.6 m
   H 26.5 m
   I 265 m

7. Where is point C on the number line?

   A 1
   B $1\frac{1}{2}$
   C $2\frac{1}{4}$
   D 2

8. The shaded region represents a courtyard surrounded by a rectangular house. What is the area of the shaded region?

   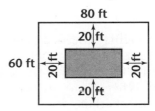

   F 1200 ft²
   G 800 ft²
   H 400 ft²
   I 200 ft²

Basic Skills Practice
Cumulative Review

Read each question and circle the best answer.

1. What is the greatest precision that can be achieved with this ruler?

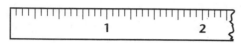

   A Nearest $\frac{1}{2}$ inch

   B Nearest $\frac{1}{4}$ inch

   C Nearest $\frac{1}{8}$ inch

   D Nearest $\frac{1}{16}$ inch

2. Rose needed to cut a 12-yard piece of ribbon into 8 equal pieces. Find the length of each piece, written as a mixed number.

   F 1 yd

   G $1\frac{1}{2}$ yd

   H $1\frac{3}{4}$ yd

   I $1\frac{11}{12}$ yd

3. Stephen earned $55 last week. He spent 9% of his earnings on entertainment. What decimal represents this percent?

   A 0.9

   B 0.09

   C 0.55

   D 5.5

4. A string of lights is needed to go around a circular stage. The radius of the stage is 10 meters. Find the length of the string of lights that is needed to surround the stage without any overlap.

   F 10π m

   G 20π m

   H 100π m

   I 200π m

5. At Morton High School, 25% of the students ride the bus to and from school. There are 558 students at the school. Find the proportion that represents B, the number of students that ride the bus to and from school.

   A $\frac{25}{100} = \frac{558}{B}$

   B $\frac{25}{100} = \frac{B}{558}$

   C $\frac{25}{558} = \frac{B}{100}$

   D $\frac{25}{B} = \frac{558}{100}$

6. The square root of which number is ***not*** between the numbers 7 and 8?

   F 48.54

   G 50.54

   H 54.54

   I 56.65

7. Joe is 21, Sean is 22, and Ben is 16. Joe is $3\frac{1}{2}$ times as old as Susan. How old is Susan?

   A 74

   B 56

   C 7

   D 6

8. Find the fraction, decimal, and percent that are equivalent.

   F $\frac{1}{2}$, 0.05, 10%

   G $\frac{1}{2}$, 0.02, 20%

   H $\frac{1}{5}$, 0.20, 20%

   I $\frac{1}{5}$, 0.02, 20%

Basic Skills Practice

Cumulative Review

Read each question and circle the best answer.

1. Using the formula provided, find the volume of a cube that is 5 inches on each side.

   $$V = s^3$$

   A 15 in.^{3}
   B 25 in.^{3}
   C 125 in.^{3}
   D 625 in.^{3}

2. Consider the expression $5x - 6y$. Evaluate the expression if $x = 4.5$ and $y = 7.8$.

   F -24.3
   G -12.0
   H -6.3
   I 25.3

3. Find the correct order from least to greatest.

   A 36,081; 36,180; 306,810; 38,016
   B 36,081; 38,016; 36,180; 306,810
   C 306,810; 38,016; 36,180; 36,081
   D 36,016; 36,180; 38,016; 306,810

4. To make a lion costume, Ann bought $3\frac{1}{4}$ yards of light brown felt. She needed to use only $2\frac{7}{8}$ yards of it. How much fabric did she have left?

   F $\frac{1}{8}$ yd

   G $\frac{2}{3}$ yd

   H $\frac{3}{8}$ yd

   I $1\frac{3}{8}$ yd

5. Monica wants to budget her money for the purchase of a new lamp. She went to several stores and made a list of the prices she found. Find the mean and the mode of the prices.

   $45.50, $45.99, $50.29, $45.50, $47.98

   A $45.50, $47.05
   B $47.05, $45.50
   C $47.05, $46.99
   D $4.79, $45.50

6. An equivalent form of 4000 mm is—

   F 0.4 m
   G 4 m
   H 40 m
   I 400 m

7. Jan bought 5 yards of fabric. She used $\frac{1}{3}$ of it to sew curtains and $\frac{1}{5}$ of it to sew a pillow. How much fabric did she have remaining?

   A 2 yd

   B $2\frac{1}{3}$ yd

   C $2\frac{2}{3}$ yd

   D 3 yd

8. Nikolette answered 78 out of 90 questions correctly on her math final. What score did she receive, written as a percent?

   F 78%
   G 82%
   H 87%
   I 90%

Basic Skills Practice

Cumulative Review

Read each question and circle the best answer.

1. Solve the equation $2x + 4 = x - 1$ for x.

   A 1
   B -1
   C 3
   D -5

2. What is the probability that the spinner will land on a number less than 10?

   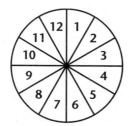

   F $\frac{5}{6}$
   G $\frac{1}{2}$
   H $\frac{2}{3}$
   I $\frac{3}{4}$

3. The ratio of girls to boys in a class is 5 to 8. If the class has 35 girls, how many boys are there?

   A 13
   B 20
   C 56
   D 64

4. An equilateral triangle has a perimeter of 36.6 centimeters. What is the length of one of its sides?

   F 9.15 cm
   G 12.2 cm
   H 18.3 cm
   I 24.4 cm

5. The temperatures for 6 consecutive days in August were 98, 100, 101, 99, 97, and 100 degrees Fahrenheit. What is the median temperature rounded to the nearest tenth?

   A 98.1°F
   B 99.2°F
   C 99.5°F
   D 100.0°F

6. What is $0.345 + 2.812$ rounded to the nearest tenth?

   F 3.1
   G 3.2
   H 3.15
   I 3.157

7. The movers charged $35.00 to move a desk within the city limits, plus $0.30 per mile. If the total charge was $36.20, not including tax, which equation could be used to find m, the number of miles that the desk was moved?

   A $0.30(35 + m) = 36.20$
   B $35m = \frac{36.20}{0.30}$
   C $35 + 0.30 = \frac{36.20}{m}$
   D $35 + 0.30m = 36.20$

8. A bag contains 20 marbles. There are 8 white marbles, 7 red marbles, and 5 blue marbles. One marble is drawn from the bag at random. What is the probability it is a blue marble?

   F $\frac{7}{20}$
   G $\frac{3}{5}$
   H $\frac{2}{5}$
   I $\frac{1}{4}$

Basic Skills Practice

Cumulative Review

Read each question and circle the letter for the best answer.

1. The volume of a cube-shaped package is 27 cubic centimeters. What is the length of one side?

   A 2 cm

   B 3 cm

   C 4 cm

   D 5 cm

2. What is $2 \times 4 + 4 - 8 \div 2 + 3^2$?

   F 11

   G 17

   H 18

   I 20

3. A recipe calls for $3\frac{1}{4}$ cups of flour. The cook has only a $\frac{1}{2}$-cup measuring cup. How many times should he fill the measuring cup to obtain the amount of flour he needs?

   A $3\frac{1}{4}$ times

   B $5\frac{1}{4}$ times

   C $5\frac{1}{2}$ times

   D $6\frac{1}{2}$ times

4. Janie had $40 to spend at a craft store. She bought a wicker basket for $10.99 and 2 frames for $5.99 each. Including the sales tax of $1.84, what was the total amount she paid for these items?

   F $15.19

   G $18.82

   H $22.97

   I $24.81

5. At a local high school, 5 out of 7 students leave campus for lunch. The approximate percent of students that leave campus for lunch is—

   A 7%

   B 70%

   C 71%

   D 74%

6. Pat bought a mountain bike for $119 two years ago. The bike is now worth $75. What is the approximate percent decrease in value?

   F 59%

   G 45%

   H 44%

   I 37%

7. Solve the equation $2x - 5 = 6x + 3$ for x.

   A -32

   B -2

   C 2

   D 32

8. Find $20 - 11.9125$ rounded to the nearest hundredth.

   F 8.08

   G 8.09

   H 8.088

   I 8.0875

9. Albert has 5 shirts and 3 pairs of slacks. How many different outfits consisting of 1 shirt and 1 pair of slacks can he make?

   A 15

   B 10

   C 9

   D 8

ANSWERS

Metric and Customary Units, Part 1, page 1

1. 6 ft   2. 360 in.   3. $\frac{1}{2}$ mi   4. 2 mi

5. 12,320 yd   6. $7\frac{1}{2}$ qt   7. 12 gal

8. 24 gal   9. 48 yd   10. 4000 m

11. 0.056 kg   12. 133,000 mL   13. 0.8 cm

14. 0.7 m   15. 3.2 g   16. 2.3 L

17. 20 mm   18. 30 cm   19. 0.037 km

20. 62 c   21. 9 mi   22. 10,880 oz

23. 470 cm   24. 2.5 mi   25. 0.075 km

26. 880 yd   27. $45\frac{3}{4}$ c

Metric and Customary Units, Part 2, page 2

1. C   2. I   3. C   4. H   5. D   6. F

7. 59,840 yd   8. 28 ft   9. 4 times

Measurement and Precision, Part 1, page 3

1. $\frac{1}{8}$ inch   2. $\frac{1}{4}$ inch   3. a   4. b   5. c

6. b   7. a   8. c

Measurement and Precision, Part 2, page 4

1. D   2. H   3. B   4. G   5. D   6. H

7. 2   8. 12.5

Estimation, Part 1, page 5

Answers may vary but should be close to the given answer.

1. 39,000   2. 6   3. 90   4. 36   5. 4600

6. 4   7. 12,000   8. 930   9. b   10. a

11. a   12. b

13. Answers may vary but should be about 28,000 lb.

Estimation, Part 2, page 6

1. B   2. H   3. D   4. H   5. C   6. I

7. $46   8. $34   9. $44

Order of Operations, Part 1, page 7

1. 14   2. 10   3. 72   4. 64   5. 36

6. 23   7. 41   8. 143   9. 24   10. 11

11. 27   12. 12   13. 22   14. 3   15. 150

16. 27   17. 48   18. 7   19. 9   20. 42

Order of Operations, Part 2, page 8

1. A   2. G   3. B   4. I   5. A   6. G

7. 28   8. 7   9. 69   10. 22   11. 27

12. 73   13. 50

Cumulative Review, page 9

1. D   2. G   3. D   4. G   5. 15   6. 5256

7. 36   8. 57   9. 10,000   10. $\frac{1}{16}$

Evaluating Expressions and Formulas, Part 1, page 10

1. 12   2. 25   3. 30   4. 21   5. −85

6. 105   7. 0   8. 213   9. 0   10. 11

11. 1   12. $\frac{1}{5}$ 13. 1   14. −3   15. 23

16. 7   17. 50°F   18. 77°F   19. 212°F

20. 32°F   21. 98.6°F   22. 122°F

ANSWERS

Evaluating Expressions and Formulas, Part 2, page 11

1. D **2.** F **3.** D **4.** F **5.** D **6.** G

7. −27 **8.** 17 **9.** 72 **10.** 37 **11.** 45

Finding Patterns, Part 1, page 12

1. 10 **2.** 13 **3.** 45 **4.** 21 **5.** 40

6. 19 **7.** 35 **8.** 42 **9.** 6 **10.** 636

11. 44 **12.** 343 **13.** dDd **14.** PrT

15. D4d **16.** 61,218 **17.** 500 **18.** 222

19. 478 **20.** 52 **21.** 44 **22.** Mnp, Qrt

23. 678 **24.** 765 **25.** 56H **26.** DW, EV

27. 81, 729 **28.** W3D **29.** 240

Finding Patterns, Part 2, page 13

1. B **2.** G **3.** B **4.** F **5.** C **6.** H

7. C **8.** F

Using Differences to Find Patterns, Part 1, page 14

1. 93 **2.** 230 **3.** 230 **4.** 138 **5.** 42

6. −7 **7.** 96 **8.** 56 **9.** 10 **10.** 176

11. 46 **12.** 18 **13.** 343 **14.** 210

15. 1182 **16.** 112 **17.** −9 **18.** 168

Using Differences to Find Patterns, Part 2, page 15

1. C **2.** I **3.** D **4.** H **5.** D **6.** H

7. B **8.** I

Cumulative Review, page 16

1. B **2.** G **3.** C **4.** H **5.** D **6.** H

7. A **8.** 7 **9.** 10 **10.** $\frac{1}{9}$

Fractions and Equivalent Fractions, Part 1, page 17

1. $\frac{1}{4}$ **2.** $\frac{3}{12}$ **3.** $\frac{1}{4}$ **4.** $\frac{3}{16}$ **5.** $\frac{2}{6}$ **6.** $\frac{3}{8}$

7. = **8.** ≠ **9.** ≠ **10.** = **11.** = **12.** =

13. 1 **14.** 14 **15.** 3 **16.** 12 **17.** 1

18. 12 **19.** 72 **20.** 85 **21.** 6

Fractions and Equivalent Fractions, Part 2, page 18

1. A **2.** G **3.** C **4.** H **5.** 45 **6.** 135

7. 39

Comparing Fractions, Part 1, page 19

1. $\frac{1}{5} < \frac{2}{5}$ **2.** $\frac{7}{10} > \frac{5}{12}$ **3.** $\frac{5}{14} < \frac{3}{5}$ **4.** <

5. = **6.** < **7.** < **8.** < **9.** > **10.** <

11. =

Comparing Fractions, Part 2, page 20

1. A **2.** H **3.** A **4.** G **5.** $\frac{3}{4}$ **6.** $\frac{1}{16}$

Mixed Numbers, Part 1, page 21

1. $\frac{5}{3}$ **2.** $\frac{9}{4}$ **3.** $\frac{16}{3}$ **4.** $\frac{7}{2}$ **5.** $\frac{66}{7}$ **6.** $\frac{19}{7}$

7. $\frac{104}{9}$ **8.** $\frac{27}{5}$ **9.** $\frac{73}{12}$ **10.** $\frac{35}{8}$ **11.** $\frac{47}{6}$

12. $\frac{90}{11}$ **13.** $2\frac{3}{4}$ **14.** $2\frac{1}{6}$ **15.** $4\frac{7}{9}$

16. $4\frac{4}{7}$ **17.** 2 **18.** $8\frac{1}{4}$ **19.** $3\frac{1}{2}$ **20.** $5\frac{1}{2}$

21. $5\frac{2}{5}$ **22.** 7 **23.** $18\frac{2}{3}$ **24.** $20\frac{1}{4}$

Mixed Numbers, Part 2, page 22

1. D **2.** I **3.** D **4.** F **5.** $862.50

6. $\frac{3}{8}$ **7.** $337.50

ANSWERS

Cumulative Review, page 23

1. A **2.** H **3.** D **4.** H **5.** $22.70

6. 80

Adding and Subtracting Fractions, Part 1, page 24

1. $\frac{7}{9}$ **2.** $6\frac{1}{10}$ **3.** $2\frac{2}{5}$ **4.** $1\frac{3}{8}$ **5.** $\frac{26}{35}$

6. $4\frac{1}{8}$ **7.** $17\frac{2}{9}$ **8.** $6\frac{7}{12}$ **9.** $9\frac{17}{24}$ **10.** $\frac{2}{11}$

11. $1\frac{4}{39}$ **12.** $\frac{7}{9}$ **13.** $\frac{3}{28}$ **14.** $\frac{13}{30}$ **15.** $\frac{1}{10}$

16. $\frac{2}{3}$ **17.** $\frac{11}{36}$ **18.** $\frac{3}{16}$ **19.** $9\frac{3}{10}$ **20.** $\frac{19}{20}$

21. $\frac{17}{24}$ **22.** $2\frac{1}{12}$ **23.** $\frac{11}{30}$ **24.** $9\frac{2}{5}$ **25.** $\frac{1}{2}$

26. $\frac{7}{30}$ **27.** $16\frac{2}{3}$

Adding and Subtracting Fractions, Part 2, page 25

1. D **2.** G **3.** A **4.** G **5.** $6\frac{3}{4}$ **6.** $16\frac{1}{2}$

7. $2\frac{11}{24}$

Multiplying and Dividing Fractions, Part 1, page 26

1. 6 **2.** $\frac{3}{7}$ **3.** $\frac{1}{33}$ **4.** $\frac{3}{25}$ **5.** 8 **6.** $\frac{1}{18}$

7. $\frac{7}{20}$ **8.** $1\frac{5}{27}$ **9.** $6\frac{1}{2}$ **10.** $\frac{1}{25}$ **11.** $\frac{9}{10}$

12. $1\frac{3}{4}$ **13.** 4 **14.** $1\frac{2}{9}$ **15.** $8\frac{2}{5}$ **16.** $\frac{7}{24}$

17. 28 **18.** $2\frac{5}{8}$ **19.** 3 **20.** $\frac{8}{11}$ **21.** 8

22. $\frac{5}{6}$ **23.** $3\frac{7}{16}$ **24.** 12 **25.** $1\frac{3}{7}$ **26.** $5\frac{1}{3}$

27. 4

Multiplying and Dividing Fractions, Part 2, page 27

1. A **2.** F **3.** B **4.** H **5.** A **6.** F

7. $1\frac{1}{3}$ or $\frac{4}{3}$

Cumulative Review, page 28

1. B **2.** H **3.** B **4.** G

5. $6\frac{1}{4}$ gal, or 25 quarts **6.** 64 **7.** $256x^{24}$

Comparing and Ordering Numbers, Part 1, page 29

1. < **2.** > **3.** < **4.** < **5.** < **6.** >

7. > **8.** < **9.** > **10.** > **11.** <

12. < **13.** 8642; 8536; 8462

14. 0.3469; 0.1452; 0.0896

15. 48,492; 48,619; 486,743

16. 0.489; 0.49; 4.89

17. Tony, Sonia, Mary, Sue

Comparing and Ordering Numbers, Part 2, page 30

1. D **2.** G **3.** C **4.** G **5.** D **6.** G

7. D **8.** F

Adding and Subtracting Decimals, Part 1, page 31

1. 8.6 **2.** 15.63 **3.** 4.379

4. 19.8 or 19.80 **5.** 67.512 **6.** 16.57

7. 29.103 **8.** 68.35 **9.** 32.75 **10.** 65.47

11. 0.2 **12.** 2.34 **13.** 3.78 **14.** 0.097

15. 0.614 **16.** 2.86 **17.** 3.339 **18.** 4.33

19. 1.93 **20.** 11.819 **21.** 20.3 in.

Adding and Subtracting Decimals, Part 2, page 32

1. A **2.** G **3.** D **4.** F **5.** C **6.** H

7. $56.30 **8.** 32.3%

ANSWERS

Cumulative Review, page 33

1. D **2.** I **3.** B **4.** G **5.** C **6.** H

7. C **8.** H **9.** 375 envelopes

10. 1120 in.^{2}

Multiplying and Dividing Decimals, Part 1, page 34

1. 5.6 **2.** 4.8 **3.** 4.5 **4.** 0.63 **5.** 19.5

6. 21.25 **7.** 30.55 **8.** 43.68 **9.** 64.328

10. 34.896 **11.** 0.7 **12.** 1.04 **13.** 24.8

14. 259 **15.** 504 **16.** 1187 **17.** 0.289

18. 83.5 **19.** 36.4 **20.** 104

Multiplying and Dividing Decimals, Part 2, page 35

1. B **2.** I **3.** C **4.** I **5.** D **6.** F

7. D **8.** I **9.** $270.60 **10.** 9

Rounding Whole Numbers and Decimals, Part 1, page 36

1. 4600 **2.** 8300 **3.** 15,200 **4.** 96,800

5. 4000 **6.** 23,000 **7.** 88,000

8. 698,000 **9.** 46 **10.** 9 **11.** 77

12. 24 **13.** 5.9 **14.** 37.3 **15.** 486.7

16. 65.40 **17.** 6.54 **18.** 45.39 **19.** 0.77

20. 16.45

Rounding Whole Numbers and Decimals, Part 2, page 37

1. A **2.** G **3.** B **4.** H **5.** B **6.** G

7. D **8.** G **9.** $45,894,000

Cumulative Review, page 38

1. B **2.** I **3.** C **4.** G **5.** C **6.** H

7. D **8.** G

Decimals and Percents, Part 1, page 39

1. 5% **2.** 22% **3.** 75% **4.** 17%

5. 41% **6.** 3% **7.** 0.82 **8.** 0.37

9. 0.07 **10.** 0.55 **11.** 0.03 **12.** 0.68

13. 54.4% **14.** 0.65% **15.** 8.5%

16. 10.5% **17.** 1.4% **18.** 175%

19. 0.034 **20.** 0.178 **21.** 1.00 **22.** 0.83

23. 0.0016 **24.** 7.552 **25.** 0.057

26. 75%

Decimals and Percents, Part 2, page 40

1. C **2.** G **3.** B **4.** G **5.** B **6.** H

7. B **8.** 0.235 **9.** 36.5% **10.** 0.60

Writing Percents as Fractions, Part 1, page 41

1. $\frac{13}{100}$ **2.** $\frac{23}{100}$ **3.** $\frac{37}{100}$ **4.** $\frac{11}{100}$ **5.** $\frac{3}{100}$

6. $\frac{99}{100}$ **7.** $\frac{49}{100}$ **8.** $\frac{7}{100}$ **9.** $\frac{39}{100}$

10. $\frac{50}{100}, \frac{1}{2}$ **11.** $\frac{75}{100}, \frac{3}{4}$ **12.** $\frac{66}{100}, \frac{33}{50}$

13. $\frac{12}{100}, \frac{3}{25}$ **14.** $\frac{24}{100}, \frac{6}{25}$ **15.** $\frac{40}{100}, \frac{2}{5}$

16. $\frac{160}{100}, \frac{8}{5}, 1\frac{3}{5}$ **17.** $\frac{175}{100}, \frac{7}{4}, 1\frac{3}{4}$

18. $\frac{120}{100}, \frac{6}{5}, 1\frac{1}{5}$ **19.** $\frac{100}{100}, \frac{1}{1}, 1$ **20.** $\frac{113}{100}, 1\frac{13}{100}$

21. $\frac{500}{100}, 5$ **22.** $\frac{75}{100}$ or $\frac{3}{4}$ **23.** $\frac{10}{100}$ or $\frac{1}{10}$

Writing Percents as Fractions, Part 2, page 42

1. B **2.** G **3.** B **4.** I **5.** B **6.** I

ANSWERS

7. $\frac{1}{2}$ 8. $\frac{3}{4}$ 9. $\frac{11}{20}$

Cumulative Review, page 43

1. D 2. I 3. B 4. F 5. A 6. G

7. B 8. G 9. 0.8 10. 1.6 11. 2

Fractions, Decimals, and Percents, Part 1, page 44

1. $\frac{45}{100}$, 0.45 2. $\frac{75}{100}$, 0.75 3. $\frac{5}{100}$, 0.05

4. $\frac{32}{100}$, 0.32 5. $\frac{80}{100}$, 0.80 6. $\frac{12}{100}$, 0.12

7. 79% 8. 82% 9. 7% 10. 26%

11. 168% 12. 11% 13. 20% 14. 30%

15. 50% 16. 16% 17. 45% 18. 34%

19. $9\frac{3}{4}$% 20. 65%

Fractions, Decimals, and Percents, Part 2, page 45

1. A 2. I 3. A 4. F 5. C 6. I

7. 0.5 8. 13% 9. 10%

Finding a Percent of a Number, Part 1, page 46

1. 7 2. 21 3. 9 4. 34 5. 93 6. 57

7. 40 8. 876 9. 46.86 10. 10.05

11. 53.7 12. 30.72 13. 65 14. 144

15. 24 16. 54 17. 32 18. 11 19. 9

20. 9 21. 108.3 22. $12 23. $1.75

Finding a Percent of a Number, Part 2, page 47

1. C 2. H 3. D 4. G 5. C 6. G

7. 51 8. 102

Cumulative Review, page 48

1. A 2. G 3. C 4. I 5. B 6. H

7. 2500 milligrams 8. -9 9. 0.0432

10. $4.50

Finding the Percent One Number Is of Another, Part 1, page 49

1. $n = 0.2 = 20\%$ 2. $n = 0.25 = 25\%$

3. $n = 0.8 = 80\%$ 4. 65% 5. 35%

6. 48% 7. 50% 8. 25% 9. 75%

10. 20% 11. 20% 12. 80%

Finding the Percent One Number Is of Another, Part 2, page 50

1. D 2. H 3. C 4. H 5. A 6. H

7. 20% 8. 40% 9. 80% 10. 80%

11. 60% 12. 55%

Finding a Number When a Percent of It Is Known, Part 1, page 51

1. 20 2. 200 3. 24 4. $60 5. $60

6. $120 7. 50 8. 175 9. 150 10. 75

11. 150 12. 30 13. 600 14. 250

15. 550 16. 900 17. 25 18. 296

Finding a Number When a Percent of It Is Known, Part 2, page 52

1. B 2. H 3. A 4. I 5. 5 6. 80

7. 22 8. 48 9. 25 mph 10. 200

11. 50 12. $100

Cumulative Review, page 53

1. B 2. H 3. B 4. I 5. B 6. H

ANSWERS

- -

7. C **8.** 300 **9.** 60% **10.** 36.1

11. 18.76

Using Proportions to Solve Percent Problems, Part 1, page 54

1. 24 **2.** 40 **3.** 3 **4.** 10% **5.** 60%

6. 50 **7.** 25 **8.** 30% **9.** 75%

10. 30 students **11.** 68 questions

Using Proportions to Solve Percent Problems, Part 2, page 55

1. C **2.** I **3.** A **4.** I **5.** C **6.** G

7. C **8.** G **9.** A **10.** 8

Percent Increase and Decrease, Part 1, page 56

1. 15%, 30%, 25% **2.** 15%, 40%, 10%

3. 80% **4.** 25% **5.** 40% **6.** 50%

7. 60%

Percent Increase and Decrease, Part 2, page 57

1. A **2.** F **3.** B **4.** I **5.** C **6.** G

7. C **8.** F **9.** C **10.** F

Cumulative Review, page 58

1. D **2.** H **3.** A **4.** I **5.** B **6.** I

7. A **8.** G **9.** 250 **10.** 20%

Using Ratios and Rates, Part 1, page 59

1. $\frac{10}{12}, \frac{15}{18}, \frac{20}{24}$ **2.** $\frac{4}{6}, \frac{8}{12}, \frac{16}{24}$ **3.** $\frac{5}{21}$ or $\frac{6}{20}$

4. $\frac{4}{12}$ or $\frac{1}{3}$

Using Ratios and Rates, Part 2, page 60

1. D **2.** H **3.** A **4.** G **5.** D **6.** H

7. C **8.** G **9.** B

Using Proportions, Part 1, page 61

1. yes **2.** no **3.** yes **4.** no **5.** yes

6. yes **7.** 84 **8.** 18 **9.** 27 **10.** 45

11. 6 **12.** 0.8 **13.** 325 calories

14. 6 rolls **15.** 266.5 miles

Using Proportions, Part 2, page 62

1. C **2.** I **3.** B **4.** I **5.** C **6.** F

7. C **8.** G

Cumulative Review, page 63

1. A **2.** I **3.** C **4.** H **5.** C **6.** H

7. D **8.** H

Points, Lines, and Planes, Part 1, page 64

1. line segment **2.** angle **3.** line **4.** ray

5. line; $\overleftrightarrow{AB}$ **6.** ray; $\overrightarrow{MA}$

7. angle; $\angle FSB$, $\angle BSF$, or $\angle S$

8. plane; plane LMN **9.** line segment; $\overline{PS}$

10. point; point R **11.** angle; $\angle V$

12. plane; plane X

Points, Lines, and Planes, Part 2, page 65

1. C **2.** F **3.** A **4.** G **5.** C **6.** C

7. I **8.** G **9.** 15 **10.** $\angle PQR$, $\angle RQP$, $\angle 3$, $\angle Q$

ANSWERS

Angles and Angle Relationships, Part 1, page 66

1. right **2.** straight **3.** acute **4.** straight

5. right **6.** supplementary **7.** neither

8. supplementary **9.** neither **10.** neither

Angles and Angle Relationships, Part 2, page 67

1. C **2.** F **3.** C **4.** F **5.** B **6.** G

7. A **8.** G **9.** 145° **10.** 63°

Parallel Lines in Geometry, Part 1, page 68

1. ∠5 **2.** ∠6 **3.** ∠7 **4.** ∠8 **5.** ∠7

6. ∠5 **7.** 150° **8.** 150° **9.** 30°

10. 150° **11.** 30° **12.** 30° **13.** 133°

14. 47° **15.** 47° **16.** 47° **17.** 47°

18. 133° **19.** 47° **20.** 133°

Parallel Lines in Geometry, Part 2, page 69

1. C **2.** H **3.** D **4.** I **5.** A **6.** H

7. D **8.** 50° **9.** 50°

Cumulative Review, page 70

1. D **2.** G **3.** D **4.** G **5.** C **6.** G

7. D **8.** $14.96 **9.** 9

Statistics: Mean and Range, Part 1, page 71

1. 85 **2.** 4.8 **3.** 49 **4.** 4 **5.** 16

6. 2.1 **7.** 25 **8.** 3.1

9. range of 25; mean of 87 **10.** 10 hours

11. 16 points **12.** 30 miles

Statistics: Mean and Range, Part 2, page 72

1. D **2.** F **3.** B **4.** G **5.** C **6.** G

7. C **8.** 21 students **9.** 29

Statistics: Mean and Mode, Part 1, page 73

1. 7 **2.** 6 **3.** mean of 5; mode of 6

4. mean of 7; mode of 5

5. mean of 73.8; no mode

6. mean of 2.9; mode of 2 and 3

7. mean of 3.5; no mode

8. 11.1 **9.** 11.2 **10.** 1.7

Statistics: Mean and Mode, Part 2, page 74

1. C **2.** F **3.** A **4.** F **5.** C **6.** G

7. C **8.** I **9.** 21

Cumulative Review, page 75

1. C **2.** F **3.** A **4.** F **5.** D **6.** H

7. B **8.** H

Squares and Square Roots, Part 1, page 76

1. 16 **2.** 49 **3.** 529 **4.** 121 **5.** 400

6. 2601 **7.** 5 **8.** 1 **9.** 9 **10.** 6

11. 10 **12.** $\frac{3}{4}$

Squares and Square Roots, Part 2, page 77

1. C **2.** G **3.** B **4.** F **5.** D **6.** I

ANSWERS

7. A **8.** G **9.** 256 **10.** 289

The "Pythagorean" Right-Triangle Theorem, Part 1, page 78

1. no **2.** yes **3.** 20 cm **4.** 8 m **5.** 39

6. 20 **7.** 6 **8.** 35

The "Pythagorean" Right-Triangle Theorem, Part 2, page 79

1. A **2.** I **3.** A **4.** I **5.** yes **6.** no

7. no **8.** no **9.** 65 **10.** 40 **11.** 16

12. 20

Cumulative Review, page 80

1. A **2.** G **3.** B **4.** H **5.** D **6.** H

7. 9.5 **8.** 30 **9.** 8

Cumulative Review, page 81

1. B **2.** H **3.** D **4.** H **5.** A **6.** H

7. $37\frac{1}{2}$ **8.** $\frac{1}{6}$ **9.** 27

Cumulative Review, page 82

1. C **2.** G **3.** D **4.** G **5.** E **6.** G

7. $4.28 **8.** 4.6 **9.** 6

Cumulative Review, page 83

1. B **2.** H **3.** C **4.** H **5.** C **6.** G

7. 125 **8.** 420

Writing Equations and Inequalities, Part 1, page 84

1. $n + 1 > 3$ **2.** $n - 2 = 7$ **3.** $10n = 50$

4. $n < 12$ **5.** $n + 30 = 64$

6. $n + 1 = \frac{24}{2}$, or $n + 1 = \frac{1}{2} \times 24$

7. $17 = 42 - \frac{1}{2}n$ **8.** $\frac{12n}{6} > 58$ **9.** $18n = 36$

10. $n + 4 < 38$ **11.** $20 \div (15 - n) = 3$

12. $8(n + 3) = 35$ **13.** $2n + 13 \le 61$

14. $5n + 4 \le 39$

Writing Equations and Inequalities, Part 2, page 85

1. B **2.** H **3.** A **4.** I **5.** A **6.** H

7. C **8.** H **9.** D

Generating Formulas, Part 1, page 86

1. $45 \le n \le 65$ **2.** 750 m^2 **3.** 36 years old

Generating Formulas, Part 2, page 87

1. D **2.** G **3.** C **4.** H **5.** C **6.** 40

7. 45 **8.** 70

Cumulative Review, page 88

1. B **2.** F **3.** C **4.** G **5.** C **6.** H

7. B **8.** 5 **9.** 0.6

Solving Equations by Addition or Subtraction, Part 1, page 89

1. 9 **2.** 9 **3.** 57 **4.** 82 **5.** 96 **6.** 983

7. 44 **8.** 30 **9.** 56 **10.** 72 **11.** 424

12. 698 **13.** 97 **14.** 154 **15.** 272

16. 7 **17.** 2097 **18.** 786

Solving Equations by Addition or Subtraction, Part 2, page 90

1. A **2.** I **3.** B **4.** F **5.** C **6.** H

7. 237 **8.** $191 **9.** $7979.76

ANSWERS

Solving Equations by Multiplication or Division, Part 1, page 91

1. 8 **2.** 17 **3.** 135 **4.** 84 **5.** 31 **6.** 3

7. 576 **8.** 120 **9.** 60 mph **10.** 24 wpm

Solving Equations by Multiplication or Division, Part 2, page 92

1. B **2.** H **3.** B **4.** F **5.** D **6.** H

7. 68 **8.** 12 **9.** 1067

Solving Two-Step Equations, Part 1, page 93

1. 2 **2.** 3 **3.** 4 **4.** 10 **5.** 10 **6.** 12

7. 16 **8.** 6 **9.** $\frac{8}{6} = \frac{4}{3}$, or $1\frac{1}{3}$ **10.** 7

11. 20 **12.** 2 **13.** 42

Solving Two-Step Equations, Part 2, page 94

1. B **2.** H **3.** B **4.** H **5.** D **6.** G

7. 85 **8.** 20 **9.** 47°

Cumulative Review, page 95

1. C **2.** F **3.** B **4.** H **5.** D **6.** H

7. 70 **8.** 25.5 **9.** 27

Graphing Points on a Line, Part 1, page 96

1.

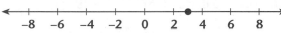

2.

3.

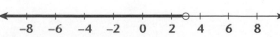

4.

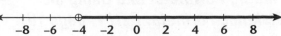

5.

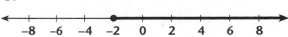

6.

7.

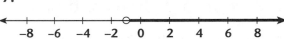

Graphing Points on a Line, Part 2, page 97

1. C **2.** F **3.** D **4.** G **5.** D **6.** H

7. $35 **8.** 65.4

Graphing Points on a Coordinate Grid, Part 1, page 98

1. $(-2, -2)$ **2.** $(2, 3)$ **3.** $(0, 5)$ **4.** $(-6, 1)$

5. $(2, -4)$ **6.** $(-3, 6)$ **7.** $(-5, -4)$

8. $(4, -3)$ **9.** *K* **10.** *I* **11.** *T* **12.** *O*

13. *S* **14.** *J* **15.** *L* **16.** *M* **17.** *R* **18.** *P*

19. *Q* **20.** *N*

Graphing Points on a Coordinate Grid, Part 2, page 99

1. C **2.** I **3.** A **4.** G **5.** C **6.** F

7. D **8.** I

Cumulative Review, page 100

1. A **2.** I **3.** C **4.** H **5.** A **6.** G

7. C **8.** I **9.** C

ANSWERS

Finding Perimeter and Using the Distance Formula, Part 1, page 101

1. 20 ft **2.** 28 yd **3.** 21 m **4.** 19 in.

5. 18 cm **6.** 16 mm **7.** 300 **8.** 144

9. 4.53 **10.** 46.83 **11.** 52.8 **12.** 0.41

Finding Perimeter and Using the Distance Formula, Part 2, page 102

1. C **2.** H **3.** A **4.** G **5.** D **6.** G

7. 260 **8.** 160 **9.** 3.37 **10.** 21.26

11. 156

Areas of Triangles, Rectangles, and Squares, Part 1, page 103

1. $17\frac{7}{8}$ in.^{2} **2.** 14.04 cm^2 **3.** $32\frac{3}{8}$ yd^2

4. 12.76 mm^2 **5.** $6\frac{1}{4}$ in.^{2} **6.** 324 mm^2

7. 60.84 cm^2 **8.** 256 in.^{2} **9.** 10.88 mm^2

10. 140 mm^2 **11.** 126.5 cm^2

12. 14.625 in.^{2}

Areas of Triangles, Rectangles, and Squares, Part 2, page 104

1. B **2.** F **3.** A **4.** F **5.** 96 **6.** 30.25

7. 17.49 **8.** 624 **9.** 344

Cumulative Review, page 105

1. C **2.** I **3.** A **4.** H **5.** D **6.** G

7. 1.2 **8.** 24

Circumference and Area of Circles, Part 1, page 106

1. 12.56 mm **2.** 21.98 in. **3.** 18.84 in.

4. 34.54 cm **5.** 28.26 mm^2 **6.** 254.34 in.^{2}

7. 78.5 mm^2 **8.** 615.44 cm^2

Circumference and Area of Circles, Part 2, page 107

1. C **2.** G **3.** A **4.** I **5.** A **6.** I

7. 397.41 **8.** 362.87 **9.** 10.05 **10.** 6.15

11. 28.26 **12.** 1519.76 **13.** 34.56

Surface Area and Volume, Part 1, page 108

1. $S = 111.0$ cm^2; $V = 77.9$ cm^3

2. $S = 34$ m^2; $V = 10$ m^3

3. $S = 105.8$ cm^2; $V = 74.1$ cm^3

4. $S = 20\pi$ in.^{2}, or 62.8 in.^{2}; $V = 20\pi$ in.^{3}, or 62.8 in.^{3}

Surface Area and Volume, Part 2, page 109

1. C **2.** I **3.** B **4.** H **5.** A **6.** I

7. approximately 88 cm^3 **8.** 56

Cumulative Review, page 110

1. C **2.** H **3.** B **4.** H **5.** C **6.** F

7. 5 **8.** $13.09

Similar Figures, Part 1, page 111

1. yes **2.** no **3.** 9 **4.** 8

Similar Figures, Part 2, page 112

1. C **2.** G **3.** B **4.** G **5.** B **6.** G

7. B **8.** H **9.** D

ANSWERS

Symmetry and Rotations, Part 1, page 113

1. vertical line of symmetry

2. no symmetry

3. horizontal line of symmetry

4. no symmetry

5. rotational symmetry, vertical symmetry, and horizontal line of symmetry

6. vertical line of symmetry

Symmetry and Rotations, Part 2, page 114

1. D 2. F 3. D 4. I 5. A 6. I

7. A 8. F

Cumulative Review, page 115

1. B 2. H 3. A 4. G 5. B 6. F

7. C 8. G

Reflections and Translations, Part 1, page 116

1. $K(3, 2)$ 2. $\triangle KLM$ 3. $S(5, -2)$ 4. $\overline{ST}$

5. $A'(4, 5), B'(9, 5), C'(9, 8), D'(4, 8)$

6. $A'(-5, 3), B'(-10, 3), C'(-10, 5), D'(1, 5)$

7. $A'(6, -4), B'(9, 0), C'(9, -6), D'(6, -6)$

8. $A'(-1, 0), B'(1, 2), C'(2, -1)$

Reflections and Translations, Part 2, page 117

1. C 2. F 3. A 4. H 5. B 6. F

7. B

Scale and Dilations, Part 1, page 118

1. 7.5 2. $56.25 \, \text{cm}^2$ 3. $6 \, \text{cm}^2$

4. $24 \, \text{cm}^2$ 5. n^2

Scale and Dilations, Part 2, page 119

1. A 2. H 3. A 4. H 5. B 6. G

7. C

Cumulative Review, page 120

1. A 2. I 3. B 4. I 5. D 6. G

7. C

Cumulative Review, page 121

1. A 2. II 3. D 4. G 5. A 6. I

7. A 8. H

Using Scientific Notation, Part 1, page 122

1. 2 2. 3 3. 5 4. -4 5. -1 6. 4

7. 2 8. -2 9. 1 10. -6

11. 4.6×10^1 12. 7.9×10^3 13. 8×10^2

14. 6.7×10^{-5} 15. 6.5×10^6

16. 6.2×10^{-1} 17. 3.7×10^{-1}

18. 5.276×10^3 19. 4.0×10^0

20. 8.56×10^2

Using Scientific Notation, Part 2, page 123

1. D 2. G 3. A 4. G 5. C 6. I

7. B 8. G 9. 45,440 10. 14,000

ANSWERS

- -

Effects of Dimensional Changes, Part 1, page 124

1. 1.5 in., 2 in., 2.5 in. **2.** $67.5\pi \text{ m}^3$

3. 96 in.^{2} **4.** 243π in.^{3} **5.** 480 in.^{3}

6. 216 in.^{3} **7.** 64 in.^{3}

Effects of Dimensional Changes, Part 2, page 125

1. B **2.** G **3.** B **4.** H **5.** C **6.** F

7. B **8.** 1:9 **9.** 1:27

Cumulative Review, page 126

1. A **2.** G **3.** C **4.** I **5.** D **6.** F

7. C **8.** G **9.** 60

Counting Outcomes and Tree Diagrams, Part 1, page 127

1. cheese soup, Greek salad, cobbler dessert
cheese soup, taco salad, custard dessert
cheese soup, taco salad, cobbler dessert
tortilla soup, Greek salad, custard dessert
tortilla soup, Greek salad, cobbler dessert
tortilla soup, taco salad, custard dessert
tortilla soup, taco salad, cobbler dessert

2. 8

Counting Outcomes and Tree Diagrams, Part 2, page 128

1. C **2.** G **3.** C **4.** G **5.** A **6.** G

Probability, Part 1, page 129

1. $\frac{1}{2}$ **2.** $\frac{1}{3}$ **3.** $\frac{1}{3}$ **4.** 0 **5.** 0 **6.** $\frac{1}{2}$

7. $\frac{1}{10}$ **8.** $\frac{2}{5}$ **9.** $\frac{1}{2}$

Probability, Part 2, page 130

1. C **2.** H **3.** D **4.** G **5.** C **6.** G

7. C

Cumulative Review, page 131

1. D **2.** I **3.** B **4.** F **5.** B **6.** G

7. −6 **8.** $\frac{6}{11}$ **9.** $\frac{1}{2}$

Cumulative Review, page 132

1. A **2.** G **3.** C **4.** H **5.** A **6.** G

7. D **8.** I **9.** 1080 **10.** 15.4

Cumulative Review, page 133

1. B **2.** G **3.** B **4.** G **5.** B **6.** I

7. C

Cumulative Review, page 134

1. D **2.** G **3.** B **4.** G **5.** A **6.** G

7. C **8.** I **9.** 0

Interpreting Bar and Circle Graphs, Part 1, page 135

1. 25 **2.** 30 **3.** 5 **4.** 90 **5.** chicken

Interpreting Bar and Circle Graphs, Part 2, page 136

1. B **2.** G **3.** C **4.** H **5.** B **6.** F

7. B **8.** G

Interpreting Line Graphs and Scatter Plots, Part 1, page 137

1. 2000 **2.** 1995 **3.** 4000 **4.** decrease

5. negative

Interpreting Line Graphs and Scatter Plots, Part 2, page 138

1. C **2.** F **3.** B **4.** I **5.** B **6.** G

ANSWERS

----

7. 50 **8.** 100

Cumulative Review, page 139

1. A **2.** I **3.** D **4.** F **5.** B **6.** 1994

7. 1992 **8.** 11,000 **9.** 1000

Interpreting Venn Diagrams and Tables, Part 1, page 140

1. 350 **2.** 80 **3.** 60 **4.** 10 **5.** 17,330 ft

6. 4505 m

Interpreting Venn Diagrams and Tables, Part 2, page 141

1. A **2.** G **3.** A **4.** I **5.** C **6.** I

7. A **8.** G

Cumulative Review, page 142

1. B **2.** H **3.** D **4.** H **5.** B **6.** I

7. 1350 **8.** 1158

Cumulative Review, page 143

1. A **2.** H **3.** C **4.** F **5.** C **6.** G

7. C **8.** F

Cumulative Review, page 144

1. B **2.** F **3.** D **4.** I **5.** D **6.** G

7. C **8.** H

Cumulative Review, page 145

1. D **2.** G **3.** B **4.** II **5.** C **6.** G

7. B **8.** G

Cumulative Review, page 146

1. D **2.** G **3.** B **4.** G **5.** B **6.** F

7. D **8.** H

Cumulative Review, page 147

1. C **2.** F **3.** D **4.** H **5.** B **6.** G

7. B **8.** H

Cumulative Review, page 148

1. D **2.** I **3.** C **4.** G **5.** C **6.** G

7. D **8.** I

Cumulative Review, page 149

1. B **2.** G **3.** D **4.** I **5.** C **6.** I

7. B **8.** G **9.** A